Business, Technology and Finance

Technology Innovation Leadership in Development

A 'Middle East' (West Asia) Perspective

BUSINESS, TECHNOLOGY AND FINANCE

Additional books in this series can be found on Nova's website under the Series tab.

Additional e-books in this series can be found on Nova's website under the e-books tab.

BUSINESS, TECHNOLOGY AND FINANCE

TECHNOLOGY INNOVATION LEADERSHIP IN DEVELOPMENT

A 'MIDDLE EAST' (WEST ASIA) PERSPECTIVE

GISELLE RAMPERSAD

AND

FAY PATEL

EDITORS

nova publishers

New York

For permission to use material from this book please contact us:
Telephone 631-231-7269; Fax 631-231-8175
Web Site: http://www.novapublishers.com

Library of Congress Cataloging-in-Publication Data

Technology innovation leadership in development : a "Middle East" (West Asia) perspective / editors, Giselle Rampersad and Fay Patel.
pages cm
Includes index.
ISBN 978-1-63117-388-2 (hardcover)
1. Information technology--Economic aspects--Middle East. 2. Technological innovations--Middle East. I. Rampersad, Giselle. II. Patel, Fay.
HC415.15.Z9I5556 2014
338'.0640956--dc23

2014004783

Published by Nova Science Publishers, Inc. † New York

CONTENTS

FOREWORD

This book presents a discussion of the effective use of technology from within developing contexts particularly the Middle East, based on the needs of developing countries and focuses specifically on the challenges they face.

The authors emphasize the role of leadership in embracing e-solutions in a vast array of government and educational platforms. The book does an excellent job in encapsulating the use of innovative technology to produce much needed change in developing contexts. More significantly, it includes vital data from various contexts in developing countries around the world, especially the Middle East (Kingdom of Saudi Arabia).

The contributing authors, from diverse backgrounds, present their principal aim in creating a basis for a change from various levels of society using innovative e-solutions, as it is applicable.

The authors, who lead change within, are presented as 'agents of change' for the good of the whole society in their own contexts.

It is at this point that I have to declare a personal salute to Rampersad and Patel, as pillars from developing countries. Indeed, this book is voiced by stakeholders and leaders of the countries involved in technology innovations.

Rampersad and Patel have thus produced a book that is not only a major scholarly analysis in the area of innovation for developing countries but also points us forward to what actually needs to be done and the possible positive impact within developing contexts, as the world changes.

The editors provide a critical base on which to make their professional judgments, and thus make a significant contribution to our work of enabling innovation leadership within developing countries to embrace the bright future ahead.

Dr. Omar Abdullah Nasseef
Dean of the English Language Institute, KAU
Formerly Head of Department of Management Information Systems,
KAU, Jeddah, Kingdom of Saudi Arabia
Vice President and Member of the Administration Council
Al Salam University, Republic of Burundi

PREFACE

'Middle East' (West Asia) perspectives on technology innovation leadership bring new voices to the technology innovation discourse in international development. The approaches of technology innovation leaders in this collection demonstrate commitment to building sustainable futures among the communities in which the authors live. There is a scarcity of technology innovation perspectives from within 'Middle East'(West Asia) perspectives and the initiatives and endeavours of the innovation leaders in this book have global appeal. Other developing community innovators are inspired to compete globally as they explore future oriented sustainable innovation solutions that are context specific. Innovation and technology leadership in areas such as e-government and e-commerce are national priorities for countries such as Saudi Arabia.

While there is a dense literature available on innovation diffusion in other developing community contexts (for example South Asia, the Asia Pacific region, Africa and Latin America), this book fills the gaps in the literature on 'Middle East' (West Asia) perspectives on innovation technology diffusion in developing community contexts as it:

- Differs by focussing on current innovation leadership from within the developing community and approaches innovation diffusion from a humanist perspective.
- Presents innovations that are critical to the sustainable futures of diverse developing communities as envisaged by the community members themselves.
- Focuses on the innovative use of information technologies through proactive leadership approaches to address innovation challenges

within a broader cross-section of societies including the Middle East. Of particular note is that contributors are approaching innovation diffusion of technology as a catalyst for change in their developing community environments. The book addresses human capability as an essential requirement in sustainable innovation diffusion initiatives.

- Presents current issues in innovation diffusion though acknowledging the 'long wave' debate and critically assessing the news waves of innovation diffusion that may not fall into the traditional patterns and trends of the 'long wave' debate. Further, it proposes approaches to innovation that are framed within social responsibility and accountability paradigms calling for comprehensive, holistic approaches that include human capability as a prerequisite for sustainable development.

The vast literature and current trends in international development and innovation diffusion in developing countries have long focused on externally driven innovations with western 'çhange agents' exporting innovations (including new ideas, technologies and machinery) to developing community contexts, making them 'passive recipients' (in the Schramm, Lerner and Rogers paradigm of the 1950s and 1960s) of innovation. Such innovation diffusion projects are aligned with international political economy agendas and corporate greed. There has been less visibility of innovation promotion from within the developing community contexts by developing communities themselves and the impact of technology innovation on developing communities is less understood from their perspectives. In this book, the contributors present an insider's view on innovation from a developing community perspective through various representations of innovation leadership. Innovation leaders from within developing community contexts have identified critical areas of innovation in the adaptation of relevant new communication technologies to enhance the quality of life of their communities so that together they can build sustainable futures.. The contributing authors (as innovation leaders) provide practical examples of innovative solutions and approaches for the effective use of information technologies for their specific developing contexts. Given the challenges in developing contexts in terms of infrastructure and systems, the role of leadership is critical in the successful use of technology and in their endeavours to counteract the challenges within.

The book focuses on the innovative approaches to technology application to uplift the quality of life among developing communities. It emphasizes the unique leadership perspectives and approaches of pioneers who boldly embrace “e-solutions” in government, commerce, education and environment. “e-Solutions” refers to the increasing use of multiple new communication technologies to innovate and produce meaningful change within developing communities. This includes demonstrating leadership through innovative e-government, e-commerce, education and environmental solutions to enhance the socio-economic and political opportunities for all members of the community. Innovative responses include a wide range of information technology integration at various levels of society. The end goal is firmly focused on improving the quality of life and on remaining connected in local and global contexts. Developing communities continue to struggle in their specific regions within their political spaces however they find innovative ways to explore and utilize information technology for their collective wellbeing. Communities facing adversity look to alternate ways to approach the challenges ahead . Contributors to this volume respond to specific grassroots issues through a critical examination and investigation of key research questions from the perspective of developing communities who are committed to building sustainable futures. In unique and creative ways, they respond to the needs of communities that they know well and who are part of their lived realities.

Themes and objectives that recur through the volume include the following:

- Enabling communities for sustainable living
- International development
- Innovation leadership and developing communities as catalysts for change
- Leading change through innovative solutions
- Commitment to sustainable futures

This book presents an opportunity to authors to present their critical contributions to international development through developing community-based lenses. Papers and reports are presented from perspectives and contexts that are marginalized and /or absent in the mainstream western dialogue about international development. We are committed to giving voice to unheard perspectives on the impact of innovation in new communication technology diffusion and to highlight the fact that given the space, the opportunity and the support, developing communities have incredible potential for taking charge of their futures and are focused on innovation, inspiring leadership and sustainability.

International development among developing communities, from a western perspective, is often perceived as a response to the need for improving the socio-economic conditions of developing communities. Initiatives are undertaken within 'missionary models' and modernity paradigms (Patel, 2012, pp. 19 -20) to help the less fortunate", to provide supporting frameworks that will benefit their socio-economic conditions. Western responses have resulted in the exportation of technology diffusion from external sources to developing communities, and in inadequate training and development of the receiving communities to use the resources effectively and to sustain their level of development.

This volume presents an embedded perspective on international development that is situated within the developing community perspective. In other words, contributors in this volume are the agents of innovation and change in their communities as they explore and lead change through their context –based, innovative solutions. As agents of change, contributors develop relevant and required innovative solutions to enhance the quality of life, enable their communities, and to move them forward in the twenty-first century by embracing change.

Developing communities know best what is appropriate for the enhancement of their quality of life. They have the potential of finding suitable, innovative solutions to build sustainable futures. This is evident in various literature (Avgerou, 1998; Baliamoune-Lutz, 2003; Manochehri et al., 2012; Patel et al., 2012; Qureshi, 2012). In this book, authors have demonstrated their intuitive response to identify appropriate e-solutions to build their communities within sustainable models.

This book will also inspire other developing communities to aspire to their highest potential in seeking creative solutions. This is evident in this volume as contributors take leadership in innovation to the next level. Patel in Chapter 1 critically reviews the status of innovation in the 21st century suggesting that international development initiatives in the new century have finally come to the realization that innovation leaders and developing communities within developing community contexts are catalysts for change, a perspective that has been absent in international development for over the last half century. It is a perspective that has been silenced and rejected by the hegemonic western paradigm of innovation. The chapter reviews the emerging trends in enabled innovation leadership. Alghamdi in Chapter 2 boldly takes on the task of challenging the status quo in Saudi Arabian government protocols and recommends enhancement in e-government. Chapter 3 by Hanafiah and Goodwin propose strategies that leaders can architect to improve increase maturity in e-government capability from mere information dissemination to two-way communication, integration, transaction to participation, thereby improving e-government performance. In Chapter 4, Fradley and Rampersad, devise innovative green information solutions for environmental monitoring. Part Two examines sustainable futures and key issues in developing infrastructures for community sustainability. Chapter 5 by Barashi investigates the impact of social media for small and medium enterprises compared to large enterprises. In Chapter 6, Alqahtani, Goodwin and de Vries explores the adoption of e-commerce services in the Kingdom of Saudi Arabia. Chapter 7 (Rampersad and Troshani) demands that governments account for their promises that high-speed broadband will in fact positively impact on their economies and communities. In Chapter 8, Alsharan, Goodwin and Rampersad devise a valuable framework for understanding the success of enterprise systems in small and medium enterprises in developing countries. In Chapter 9, Patel argues that a holistic framework for innovation leadership is a social responsibility imperative. Chapter 10 concludes with a review of the contributing chapters. Rampersad and Patel highlight that developing communities have taken ownership of innovation from within and are indeed the catalysts for change through strategic and negotiated blending of their indigenous knowledge with digital innovations to enhance their quality of life.

Giselle Rampersad
Fay Patel

References

Avgerou, C. (1998). How can IT enable economic growth in developing countries? *Information Technology for Development,* 8(1), 15-28.

Baliamoune-Lutz, M. (2003). An analysis of the determinants and effects of ICT diffusion in developing countries. *Information Technology for Development,* 10(3), 151-169.

Manochehri, N., Al-Esmail, R., & Ashrafi, R. (2012). Examining the Impact of Information and Communication Technologies (ICT) on Enterprise Practices: A Preliminary Perspective from Qatar. *The Electronic Journal of Information Systems in Developing Countries,* 51.

Patel, F., Sooknanan, P., Rampersad, G., & Mundkur, A. (2012). *Information Technology, Development, and Social Change*. New York: Routledge.

Qureshi, S. (2012). As the global digital divide narrows, who is being left behind? *Information Technology for Development,* 18(4), 277-280.

In: Technology Innovation Leadership … ISBN: 978-1-63117-388-2
Editors: G. Rampersad and F. Patel

Chapter 1

21ST CENTURY SUSTAINABLE INNOVATION DIFFUSION: ENABLING INNOVATION LEADERS AND DEVELOPING COMMUNITIES AS CATALYSTS FOR CHANGE

Fay Patel*

Bond University, Gold Coast, Queensland Australia

ABSTRACT

In the face of adversity, developing communities and innovation leaders explore multiple possibilities to expand the reach of innovative e-solutions. It is a challenging task for developing communities to embrace sustainable e-solutions especially when it requires a blend of the wealth of their indigenous knowledge along with the differing cultural perspectives of developed communities. In order to drive innovation from within there is an imperative to respectfully recognize the roles of innovation leaders and developing communities as catalysts for change. Further, approaches to innovation should be enabling and responsive within a sustainable paradigm. This chapter focuses on the emerging proactive initiatives in the 21st century to enable innovation leaders and developing communities as catalysts for change.

* Email: dr.fay.patel@gmail.com.

INTRODUCTION

Innovation diffusion has been approached from multiple perspectives. Various theories have been presented as to what innovation diffusion is, how it is approached, and what the general and specific trends and patterns are in the adoption of innovations by global communities. Although the innovation diffusion literature is exhaustive in a wide range of areas (Tidd, 2010; Rogers, 2003; Hall & Preston, 1988) there has been little consensus as to its definition and nearly a century later we have a wide range of definitions of innovation. Hall & Preston (1988, p.22) suggest that definitions of innovation are sometimes focused on 'product or process or other related factors such as new markets' and so on. With reference to Rogers perspective on innovation diffusion as the epidemic S-curve model, Tidd (2010. P.14) states that 'it assumes a homogenous population of potential adopters' who are 'similar and have the same needs'. That assumption is problematic. It is encouraging to observe however that in the 21st century innovation diffusion approaches are culturally diverse and context driven. These approaches align with the ongoing evolution of technological and other innovations that must respond appropriately, adequately and with immediacy to glocal (local and global) needs of developing communities. Developing communities reside within a complex demographic glocal sphere, a sphere in which local merges with global.

With regard to related theories that technology diffusion occurred in 'waves'(Kondratieff, 1935; Schumpeter, 1934), referring to the ' revival of the long wave theory and long wave debate' suggesting that there was a correlation between the 'long waves' of technology innovation diffusion and economic conditions (Hall & Preston, 1988, p.4) over the centuries. As noted by Hall and Preston, Kondratieff contented that 'long waves [....] were accompanied in changes in technology, but were not triggered by them; for new techniques can be exploited only when the economic conditions are right', Hall & Preston's (p.4) found that there was a counter argument put forth by Schumpeter. Schumpeter gave the 'long wave' theory the label Kondratieff associating his name with each new wave of innovations or economic development. According to Hall & Preston's (1988, p.4) study, Schumpeter argued that contrary to Kondratieff's contention, each new wave was in fact triggered by the new technologies and that each wave 'was indeed caused by a bunch of new innovations which created whole new industries' citing the emergence of 'coal, and steam between 1785 and 1842, steel, railways, steamships and machine tools between 1843 and 1897 and cars, electrical goods and chemicals since 1898'. Although Hall & Preston pose a relevant question (p.4, "does innovation really peak at approximately half-century intervals?) based on the patterns and analysis of technology innovations and diffusion coinciding with economic development at particular times in a century, they also claim (p,22) that there is 'no single, agreed , theory of technological change and economical development'.

However, an interesting outcome of Hall & Preston's (p.18) study of relevance to this discussion and to this century is the prediction that 'the fifth Kondratieff (or long wave of innovations) would begin sometime around the year 2010. This coming at a time when the capitalist world was just descending into the worst recession for 50 years, could hardly fail to make an intellectual impact'. Perhaps this answers Hall and Preston's question about innovation peaking at half-century intervals. However it also raises pertinent questions about why the 'fifth Kondradtieff wave' in the twenty-first century does not fit the patterns of past centuries; how the Great Recession of 2008 (Nayyar, 2011) came about before the turn of half a century, and more importantly, how the fifth wave of innovations along with the economic downturn of 2008 -2010 impacts developing communities. Nayyar (2011, p.20) contends that 'even if cycles are embedded in the nature of capitalism, it is clear that this Great Recession [of 2008] is the deepest crisis in capitalism since the Great Depression more than 75 years ago.[…] Developing countries were not as connected with the world economy at that time.[…]'. Further, Nayyar asserts that with the current integration of world economies to developing countries 'meaningful analysis of crisis and recovery must also focus on the developing world …'.

Against the aforementioned brief overview and context of innovation diffusion and economic trends over the centuries, the discussion refocuses on the emergence of a range of information and communication technology (ICTs) innovations in the twenty-first century. Notwithstanding the range of complexities that are connected to ICTs, these innovations have also brought opportunities to developing communities to embrace e-solutions or technology embedded solutions to enhance their quality of life. Within a positively entrenched paradigm of technology diffusion for empowering developing communities (Rampersad, 2012, p.25), a new breed of innovation leaders have emerged from within developing communities as catalysts of change. Innovation leadership from within developing communities endeavours to improve the socio-economic, political and environmental well-being of their host communities. This is a refreshingly positive way forward for developing communities who are led by innovative leaders in their community to embrace e-solutions that are deemed fit by those who live among them. It is a refreshing and positive take on previous approaches to innovation diffusion in international development in which innovation diffusion was driven from the outside (external sources) without regard for developing community needs and without encouraging ownership of and engagement with the innovation among the communities who were expected to embrace them. This chapter is concerned with recognising and enabling developing communities and innovation leaders as catalysts for change in their own right, and supporting innovation leaders to embed e-solutions in sustainable ways. Terminology is clarified next as related to the discussion on innovation leadership.

CLARIFICATION OF TERMINOLOGY

Within the context of this book, a wide range of perspectives are exchanged through various concepts on the subject of innovation diffusion and its impact on developing communities. Terms are clarified as they are used in this chapter to facilitate discussion and understanding of the critical issues emerging within a twenty-first century diffusion of innovation paradigm.

Innovation diffusion in this chapter subscribes to Auberg's (2005, p. 6) broad description of innovation within which technological and non-technological approaches, perspectives, products and services are included. Tidd (2010, p.7) elaborates further that it may refer to the 'diffusion of a certain technology or product but an innovation may also be also be an idea, information, belief or practice' and that (p. v) 'diffusion, or the widespread adoption of innovation is critical, but under-researched and ill-understood. It is the means by which innovations - technological, commercial and organisational – are translated into social and economic benefits'. Tidd observes (p.v) that 'most studies are concerned only with the rate of adoption or the final proportion of a population that adopts an innovation' and identifies (p.4) '…two contemporary issues in the understanding and management of the diffusion of innovations: dealing with risk and uncertainty, such as unintended consequences of the adoption or non –adoption through experimentation and learning; and the central role of networks in the diffusion and evolution of innovations'.

Innovation leadership refers to the leadership initiatives taken by members of developing communities through exploring e-solutions to respond to socio-economic, political and environmental concerns. Innovation leaders have taken responsibility to explore the possibilities and range of e-solution choices and have identified reasonable, appropriate and acceptable e-solutions that are suited to the immediate and long term needs of their developing communities.

E-solutions or eco-solutions (electronic and ecologically friendly solutions that protect the environment from harm) as *innovative solutions* refer to the electronic application of emerging information and communication technologies to promote sustainable quality of life among developing communities. *e-Solutions*, as noted earlier and within chapter discussions, also refers to the increasing use of multiple new communication technologies to innovate and produce meaningful change within developing communities. *Sustainable e-solutions* within a development paradigm refers to 'human- and resource-sensitive approach to development so that human and natural resources are not exploited and manipulated in a way that is disadvantageous to the quality of human life of a community, in the present and the future' (Patel et al, 2012, p.18).

Another term of reference that describes commitment to sustainable development in the literature is eco-innovations (Ockwell et al, 2010). Ockwell et al (2010, p.7) adopt the definition as outlined by the Environmental Technology Action Plan of the European Commission and refers to 'new production processes, new products or services, and new management and business methods, whose use or implementation is likely to prevent or substantially reduce the risks for the environment, pollution and other negative impacts of resources use, throughout the life cycle of related activities'.

Developing communities (Patel et al, 2012, p.11) 'are characterized by low socio-economic status, illiteracy levels, high level of unemployment, social inequities, lack of and inadequate access to education and health…' and is the preferred term of reference in this chapter. A diverse range of terms of reference as noted in the literature are applied from different perspectives to describe developing communities. For example, Auberg (2005, p.5) uses the term 'developing countries' to refer to 'low and medium income countries in the World Bank categorization.' Ockwell (2010, p.7) use the term 'developing countries'in their report 'for ease of use'[…]to refer to both developing and emerging economies'.

Catalysts for change describes individuals (such as innovation leaders), groups (developing communities) and events that lead change and encourage alternate perspectives and approaches in enhancing the quality of life. Auberg (2005, p.7) cites two events (the globalization process and technological change) as 'global drivers' of change. Other variations and terms of reference in the literature that refer to different aspects of leadership within other contexts are ‘ opinion leaders’, ‘change agents’, and ‘ community leaders’. Some of the literature makes a distinction between these terms of reference. For example, Tidd (2010, p.10) suggests that ‘change agents’ are the linkage (external source) between the innovation project coordinator and the ‘opinion leader’ (internal source) within the broader social community in which the innovation in being implemented. He contends that the ‘mistake made by change agents is to choose opinion leaders who are too innovative compared to the social system making the opinion leaders too atypical to act as a model and to promote change’ According to Tidd (2010, p.7) the emphasis in the Rogers’ approach to diffusion was ‘on opinion leaders and change agents working within social structures and systems’. Tidd (2010, p.10) claims that ‘opinion leaders carry information across boundaries between groups, much like knowledge bridges’. They operate from the edge of groups, rather than from the top; they are not leaders within a group, but brokers between groups’. Gaventa (2004, p.14) on the other hand, suggests that the identification of leaders within the respective community may be also be a complex process. He claims that ‘the procedures for identifying leaders can be democratic, transparent and relatively representative; often they are not. Leaders can be chosen through democratic elections; nomination by community groups; self-selection; or co-option by councillors or other partners.’’ Within the context of this book, contributors may either be self-selected or have allegiances to community or local, regional and national level governance structures. In some ways, they are self-selected leaders from within their communities and in other ways, by their defined roles and responsibilities as professionals they are also in partnership with government-led projects. Gaventa (2004, p.17) contends that ‘not only do community leaders play different roles, but they may be expected to play them differently, depending on the cultural norms of the community they represent’.

It is encouraging to note that the literature of the last decade has recognised that 21st century innovation diffusion will be driven from within developing communities with them leading innovations that are sustainable for their communities. Further, Ockwell (2010, p.7) asserts that there should be recognition ‘of the central need to foster the development of indigenous eco-innovation capabilities in developing countries’ and Gaventa (2004, p.26) contends that ‘attention must be given to strengthening the capacity of local leaders to exercise voice; at the same time, voice without responsiveness’ would be a futile strategy as it would be counterproductive. Gaventa makes an important point about embedding authentic enabling and empowering strategies among developing communities.

THE 21ST CENTURY INNOVATION PARADIGM

Current literature and international development initiatives suggest that the 21st century innovation paradigm is compassionate and people-oriented focusing on building and nurturing human capability with commitment to social responsibility and justice for sustainable development. The language of innovation diffusion is changing from an 'export model' of innovation diffusion from the Western world (Patel et al, 2012; Ockwell, 2010; Gaventa, 2004) to developing a 'domestic or local model' of innovation diffusion which is referred to as 'grassroots development' or as Auberg (2005, p.2) refers to it as responding to the 'grass-root needs at the local level' which is labelled 'the new localism'in the UK (Gaventa, 2004, p.1). Gaventa (2004, p.1) notes that the new emphasis on community participation 'in the policy decisions which affect their lives and in the design and implementation of services, especially at the local level' also focuses on '...[initiatives that are] enhancing civic life, deepening democratic participation, and contributing to more effective neighbourhood renewal and sustainable communities'.

According to the Background Note of the Africa & the ICDs Panel Report (2013, p.3), 'new technologies and innovations, in spite of their general purpose nature, are making demonstrable impact on peoples lives and welfare'. Further, the report claims that ' other factors defining the 21st century context of international cooperation with implication for technology development, transfer and diffusion include [...]...demand for greater participation and voice in global governance...'. This is an encouraging trend in international development that illustrates a new dimension to innovation diffusion with developing communities striving for ownership of the innovation diffusion global agendas. Developing communities are pushing back on past innovation diffusion models that had 'innovation diffusion done to them'.

Although the language of innovation diffusion and international development is moving toward a more inclusive, human capability orientation there is some way to go before we see significant shift toward a people-centred paradigm of innovation diffusion. For now the international literature is scattered with various combinations of people-oriented and marketing or corporate language, depending on the context of the publications. Whereas human capability embraces an enabling commitment to building capabilities (with added focus on affective and psychomotor skills) for sustainable futures and moves beyond intellectual capacity, human capital focuses on intellectual capacity (Messinis & Ahmed, 2010).

Messinis & Ahmed (2010, p.2) contend that (p.2) 'economic theory considers human capital to be the engine of economic growth' suggesting that the human factor is a key driver of economic change. Human capital is adopted in the literature (Messinis & Ahmed, 2010, p.9) to emphasize 'cognitive skills acquired at school, cognitive skills that are useful in scientific research, and cognitive skills [...] useful in the application of modern technology'. They adopt the OECD (1998) definition of the concept human capital "as the 'knowledge, skills, competencies and other attributes' that are relevant to economic activity (p.2)". Messinis & Ahmed assert that it is not simply the acquisition of formal education but that it is the application of cognitive skills or 'valuable'skills that impact technology diffusion. From their perspective, technology diffusion and adoption is not dependent on level of formal education but more specifically it is connected to the application of cognitive skills by the adult population.

Messinis & Ahmed review and examine economic theories (Schumpeter, 1934; Benhabib & Spiegel, 1994) that connect human capital to economic growth. The following assumptions about economic growth and human capital are noted among others:

- Investment in human capital (knowledge, skills, competencies and other attributes) is required for economic growth
- The level or depth of human capital affects the rate of technological change

Messinis & Ahmed (2010, p.4) examine various assumptions and 'develop a new measure of human capital that incorporates several dimensions of human capital and emphasises the application of cognitive skills by the adult population that we call 'valuable skills'.' Their study (p.19) presents 'evidence that the new 'valuable skills' index best explains trends in technology growth since the 1970s' and they 'conclude that 'valuable skills facilitate innovation and technology diffusion'. Other significant outcomes of their study are listed below.

- More countries are prone to poverty traps of which many have not caught up over three decades.
- Investment in education among African and other OECD countries (Sweden, Canada and France) have not enhanced valuable (employable) skills and hence have not facilitated innovation and technology diffusion (p.12).

- Asian and South European countries invested in valuable skills and have made considerable progress in innovation and technology diffusion.

So have we come closer to a human capability oriented innovation diffusion among developing communities that is enabling and sustainable? It may be too early in the century to tell however we should acknowledge the critical changes that have occurred in regard to the recognition of human drivers of change from within a grass-root innovation diffusion perspective. Human drivers of change or innovation leaders as catalysts for change have always been part of innovation development models as 'change agents' and 'opinion leaders' over the past century however they have not been recognised as key drivers of change from within developing communities. Within the Rogers (2005) paradigm, change agents were identified as external mechanisms to drive change from the outside whereas opinion leaders were identified within developing communities as internal 'change agents' to carry innovation forward within the community. Their role as opinion leaders was always subsumed within the export model of innovation diffusion. Within that model local, grass-root 'change agents' were identified as 'opinion leaders' to do the 'legwork' in diffusing the innovation. Various applications of the innovation model either regard change agents and opinion leaders as two distinct roles (Tuan, Singleton, Dzung & Palis, 2010) or as the same (Lam & Schaubroeck, 2000). Traditional innovation diffusion models exploited change agents and innovation leaders. However, the 21st century paradigm of innovation leadership is an empowering model in which innovation leaders self-select their role and self-identify. They promote their innovation leadership within a grass-root developing community as part of their social commitment and responsibility to their communities. Within a 21st century paradigm, innovation leaders (and developing communities) are recognised as catalysts for change. Innovation leaders are embedded within the developing community and work *with* the community to enhance their quality of life from within.

Innovation Leaders (and Developing Communities) as Catalysts for Change

Innovation leaders as catalysts for change work *with and from within* the developing community, self-select their role and self-identify, as noted earlier. They explore innovations that will bring improved quality of life in any shape or form and embark on research and study programs to educate and inform themselves in areas within which they can make a difference. It is imperative to begin recognising innovation leaders within developing community contexts as catalysts for change and as individuals who have the ability to effect change from within. They require support and enabling strategies to facilitate change in their roles. The World Bank report (Aubert, 2005) contends that 'the promotion of innovation should be considered in a gradual manner in building upon resources and capabilities available in countries at their level of development and in taking due account of their specificities including their conditions of governance'. This is a refreshing perspective on innovation diffusion since the human capabilities and resources within the developing community are being recognised as significant to the diffusion of innovations process. This new perspective is further expanded by Ockwell et al (2010, p.8) who contend that there was a need to 'recognise the central need to foster the development of indigenous eco-innovation capabilities in developing countries 'and that they should be 'producers and eco-innovators in their own right' instead of 'consumers of eco-innovations'. Further, they assert that 'policy also needs to be improved to better respond to the context-specific technological and cultural requirements which vary inter- and intra-nationally'. This new innovation diffusion approach places developing communities at the centre of innovation diffusion within an empowering paradigm that is expansive and responsive. Within such a paradigm, developing communities and their innovation leaders, will set their own agendas for change. They will shape and determine the kind of eco-innovations needed and the graduated levels of innovation diffusion through planned human capability building programs.

Within an empowered innovation diffusion model, developing communities build capability through a comprehensive framework of technology transfer strategies, as noted in Ockwell et al (2010, p.9). The framework of technology transfer adopted by Ockwell et al (2010, p.10) includes: the traditional notion of 'hardware' (physical equipment); the often ignored, yet vital, 'software' element of technology (knowledge and processes), including both underlying knowledge (e.g., engineering and manufacturing processes, or new farming techniques) and tacit knowledge (knowledge acquired by doing, e.g., applied engineering & systems integration skills). The software aspect is often most important'.

So if innovation leaders are catalysts for change in a new century, how are they enabled to impact sustainable development from within developing community contexts?

ENABLING INNOVATION LEADERSHIP WITHIN DEVELOPING COMMUNITIES

As noted earlier, innovation leaders may self-select and this is a common trait of community leaders, as suggested by Gaventa (2004, p.15), 'community leaders also emerge through self-selection'. Within the report, Gaventa (p.13) define 'community leaders' as those leaders who emerge in relationship to external partnerships with government'. Innovation leaders are not necessarily in partnership with government: they may be in government themselves or among the community and they demonstrate leadership through their identification of an innovative solution to enable their local community and to raise the quality of life in sustainable ways. They may have some things in common with other traditional notions of community leaders' such as

- the voluntary nature of their self-selected leadership role
- representation of community perspectives
- lobbying on behalf of the community at different levels of innovation diffusion within their organisation

In an in-depth discussion about participatory governance, Gaventa (2004) raises key questions about empowerment of leaders and communities, selection of community leaders, legitimacy, representation of multiple publics, and accountability within local community contexts. He contends that 'one of the assumptions of participatory forms of governance and development is that greater participation' is connected to inclusive decision- making and 'better decisions'. However, he points out that this is a flawed perspective as it leads to unclear boundaries with regard to 'traditional roles and accountabilities' (p.9) and assumes that there is a correlation between community participation and representation. Within that flawed perspective is another interpretation of empowerment that links increased visibility of leaders to empowerment. Increased visibility of leaders (regardless of the emphasis on inclusivity and their defining characteristics of representativeness such as gender and race) do not translate to empowerment. Critical issues that emerge when such assumptions are made relate to accountability and representativeness, legitimacy, whom they represent and how they balance ''insider' and 'outsider' roles' (p.10).

CONCLUSION

Enabling innovation leaders within developing communities requires support mechanisms that welcome them, create spaces for engagement with leaders, and identify clear roles and responsibilities. Strategic planning and implementation of effective mechanisms at various levels of national, organizational and local community engagement is required to ensure that there is alignment between policy and practice. If enabling policies and practices are absent, then there arises a need to establish such policy and practice, as necessary. An 'empowering communities' innovation paradigm promotes a holistic multi-level framework for innovation at various complex dimensions (at national, industry, organizational and individual levels) of innovation diffusion as advocated by Rampersad (2012, p.26) and requires a flexible, receptive innovation approach (Gaventa, 2004; Aubert, 2005). With reference to the complex dimensions of innovation, Rampersad (2012, p.17) asserts that 'more attention is required at the national level to foster an enabling culture'; communication efficiency, trust and commitment (p. 29) are necessary at a network/industry level; an innovative organizational culture that rewards innovative ideas and captures ideas through various enabling mechanisms fosters innovation (p.30); and recognising talent and providing incentives at an individual level develops and retains talent to positively impact the national economy (pp.31 -32).

This chapter critically reviewed the multiple complex dimensions of enabling innovation leaders and developing communities to take ownership of innovation from within their developing community context and highlights the need to establish enabling mechanisms to support them. The chapter draws on useful recommendations from studies by Gaventa, Ockwell and Aubert, among others, who present proactive, enabling policies and practices that will benefit all communities.

Among the practices and policies that are useful in promoting sustainable communities (developing and developed), Gaventa (p.19), citing the experience of other countries, recommends enabling measures that include those listed below. These enabling strategies may be adapted and applied to building capability in innovation leadership, as appropriate within a context-driven perspective.

- 'Strengthening voice on the one hand, to ways of strengthening receptivity to voice'
  - This requires 'creating the pre-conditions for voice, through awareness-raising and building the capacity to mobilise'.
- 'strengthened by initiatives that strengthen the receptivity to voice'
  - This requires accountability, responsiveness and recognition of the right to express voice
- Exploring new forms of governance through newly formed joint representation of both civil society and government representatives
- Allocation of funds for capacity building and training for effective participation or 'participatory budgeting' (p.20)
- Create a system of 'empowered government representatives' to hold government accountable through 'Vigilance Committees' (p.21) for example
- 'Creating forms of direct citizen participation' within local level planning for economic development, social justice, health and so on through local village assemblies and councils
- Inclusion of previously marginalized groups, for example, women

Gaventa (p.26) is emphatic that 'the development of strong effective community leadership is a process of community building, not simply a process of supporting individual leaders' (p.26). Furthermore, he asserts that new 'rules and roles of engagement' that incorporate new rights, responsibilities and accountabilities must also be developed (pp 27-28). Pre-conditions should also be established to define power differentials and to enable community leadership to be aware of rights, be able 'to mobilize, act collectively, and have the ability to communicate' with more powerful stakeholders. In conclusion, Gaventa's advice on developing empowering participatory leadership should be strongly considered. More importantly, the new paradigm should embrace the notion that 'helping communities design and develop context-appropriate indicators for measuring their own performance in the areas of community leadership, political representation and the relationship between the two might be a useful starting point'.

REFERENCES

Aubert, J-E (2005) Promoting Innovation in Developing Countries: A conceptual framework *World Bank Institute Contributions by Kuznetzov,* Y., Whyte, R., & Quirke, A.

Africa and IDCs Panel .(2013). *ECOSOC Panel Background Note.*

Benhabib, J. and Spiegel, M. (2005), Human capital and technology diffusion, in Aghion, P. and Durlauf, S. N. (eds.), *Handbook of economic growth, 1A*, 935-966, Handbooks in economics, *22*, Amsterdam and San Diego: Elsevier, North-Holland.

Butler, J.B. & Gibson, D.V. (Eds) (2011). Global Perspectives on Technology Transfer and Commercialization: *Building Innovative Ecosystems Austin,* TX, USA: Edward Elgar Publishers.

Gaventa, J. (2004). *Representation, community leadership, and participation: Citizen involvement in neighbourhood renewal and local governance* Institute of Development Studies Prepared for the Neighbourhood Renewal Unit, Office of the Deputy Prime Minister, UK.

Hall,P. and Preston, P. (1988). *The Carrier Wave - New Information Technology and the Geography of Innovation 1846-2003* London, UK: Unwin Hyman.

Jones, B. and Miller, B . (2007). *Innovation diffusion in the new economy.* The tacit component Oxon, UK: Routledge.

Kondratiev, Nicholai (also known as Kondratieff).(1925) *The major economic cycles.*

Kondratieff, N. (1935). *The long waves in economic life* Kessinger Publishing.

Kondratieff, N. & Daniels, G. (1984). *The long wave cycle.* E.P.Dutton Publisher.

Lam, Simon S. K.; Schaubroeck, John . (2000). A field experiment testing frontline opinion leaders as change agents. *Journal of Applied Psychology,* Vol 85(6), Dec 2000, 987-995. doi: 10.1037/0021-9010.85.6.987.

Mayer, Jorg. (2001). *Technology diffusion, human capital and economic growth in developing countries No. 154* United Nations Conference on Trade and Development (UNCAD) Discussion Papers.

Messinis, G. & Ahmed, A.D. (2010). *Innovation, Technology Diffusion and Poverty Traps: The role of valuable skills* Working Paper Number 47 Centre for Strategic Economic Studies, Victoria University, Melbourne Australia

Nayyar, D. (2011) The financial crisis, the Great Recession, and the Developing world *Global Policy* Volume 2 Issue 2 January 2011.

OECD .(1998). *Definition of human capital.*

Ockwell, D., Watson, J., Mallet, A., Haum.R., MacKerron, G. & Verbeken, A-M. (2010), *"Enhancing Developing Country Access to Eco-Innovation: The Case of Technology Transfer and Climate Change in a Post-2012 Policy.*

Framework", *OECD Environment Working Papers*, No. 12, OECD Publishing. http://dx.doi.org/10.1787/5kmfplm8xxf5-en.

Patel, F., Sooknanan, P., Rampersad, G., and Mundkur, A. (2012) *Information Technology, Development, and Social Change*, New York: Routledge. ISBN 13: 9780415502689

Patnaik. L.M., Talukder, A.K., Bhattaria, D.,Pradhan, H.M.. Jha, S., Iyengar, S. (2007). *Innovative Applications of Information Technology for the Developing World: Proceedings of the 3rd Asian Applied Computing Conference* London, UK: Imperial College Press

Rampersad, G. (2012). Empowering Communities A holistic approach for innovation In Patel, F., Sooknanan, P., Rampersad, G., & Mundkur, A. (2012). *Information Technology, Development, and Social Change*. New York: Routledge.

Rogers, E. M. (2005) *Diffusion of innovations*, 5th Edition, New York: Free Press.

Schumpeter, J. (1934) *The theory of economic development. An inquiry into profits, capital, credit,interest, and the business cycle*. Cambridge: Harvard University Press.

Tuan, L.A., Singleton,G.R., Dzung,N.V. & Palis, F.G. (2010). The role of change agents and opinion leaders in the diffusion of agricultural technologies in Vietnam: A case study of ACIER *–World Vision collaborative adaptive research projects* pp 261 -274.

Tidd, J. (Ed.). (2010). *Gaining momentum Managing the diffusions of innovations 2010* Singapore: Imperial College Press.

In: Technology Innovation Leadership … ISBN: 978-1-63117-388-2
Editors: G. Rampersad and F. Patel

Chapter 2

READY OR NOT? ARE SAUDI ARABIAN ORGANIZATIONS READY FOR e-GOVERNMENT?

Ibrahim Alghamdi*
Flinders University, Adelaide, Australia

ABSTRACT

This chapter discusses e-government in the Kingdom of Saudi Arabia with a particular focus on the level of e-readiness in the public sector. E-government is embryonic in Saudi Arabia and consequently, respective organizations require guidelines to establish effective e-government concepts and measures. The primary step in advancing e-government is conducting an e-readiness assessment. An e-readiness assessment aims to lead development efforts by providing standards for evaluation and progress monitoring.

The main purpose of this chapter is to investigate e-readiness assessment models and to explore the receptivity towards e-government in government organizations in the Kingdom of Saudi Arabia. Three main factors (e-services, ICT infrastructure and human skills) have been used. These factors were derived from the Center for International Development and Harvard University (CID), the Asian Pacific Economic Cooperation (APEC) models, and suggested factors by the research team of Yesser which is the national e-government program in Saudi Arabia.

* algh0044@flinders.edu.au

The results show that more efforts are fundamentally required to advance e-government in the public sector particularly regarding e-services and human skills.

INTRODUCTION

During the last 15 years, e-government has been constructed on the basis of e-government through specific technology such as m-government (mobile government), u-government (ubiquitous government), g-government (geographic government for instance through global positioning systems (GPS) applications for e-government), and t-government (transformational government). Despite the growing sophistication of e-government, as the digital divide deepens between developed countries and developing countries (Patel et al., 2012), a core issue remains whether government organizations in developing countries are ready for e-government. Defined as *"the use of ICT, and specifically the Internet, as a tool to achieve better government"* (OECD, 2003), e-government promises many benefits to the developing world including faster, streamlined administrative processing; lower transaction costs; better use of information resources; greater public access to government information and services; and more opportunities for public participation in democratic processes (O'Neill, 2009).

The main purpose of this chapter is to identify and analyse the primary issues, opportunities and challenges that e-government initiatives face in the Kingdom of Saudi Arabia.

COUNTRY PROFILE

Saudi Arabia has embarked on an e-government initiative to advance the digital economy for its citizens. To achieve this vision, Saudi Arabia has taken steps to develop business processes and spread the concept of e-services in different government organizations (Yesser, 2013). The Government of Saudi Arabia promotes the significant advantages that e-government can provide to the national economy. Thus, the supreme Royal Decree number: 7/B/33181, dated 7/9/2003 included a directive to the Ministry of Communications and Information Technology to prepare a national e-government plan in order to provide electronic services and transactions (Yesser, 2013).

To achieve the above objectives, Saudi Arabia adopted a set of promising motivated plans and strategies. Transformation to an information society cannot be gained without inclusive collaboration and intensive attempts to implement the plan's objectives. Consequently, the Ministry of Communications and Information Technology (MCIT) established the e-Government Program "Yesser" in 2005 in combination with the Ministry of Finance and the Communication and Information Technology Commission (CITC) (Yesser, 2013). Yesser has the responsibility for developing and implementing these plans and strategies in cooperation with public organizations.

Yesser aims to advance the development and performance of employees in the public sector and underline achievements as they create the perception of e-services in Saudi Arabian society. The vision of the e-government initiative in Saudi Arabia is designed to encourage and to be a channel for government entities in implementing e-services for enhancing and serving the community (Yesser, 2013).

The main objectives of Yesser are to: 1) assist government agencies in providing better and easier-to-use services for individual and business customers, 2) inspire government agencies to work together for providing integrated electronic services with a higher efficiency for individuals and businesses, 3) highlight and value the extraordinary efforts provided by the e-services and e-government teams within government agencies, 4) encourage government agencies to participate in developing the National economy by providing more easy-to-use e-services to businesses, 5) support a culture of "best standards and practices" and encourage government agencies to adopt "top-standards" in e-services, and open doors to include the public for helping to design better e-government services.

In order to achieve these objectives, the e-government vision relies on formulating a Saudi Arabian e-government strategy based on using ICT to improve government readiness

Yesser created channels to facilitate this vision which are Government Security Network (GSN) and Government Service Bus (GSB) in addition to the Saudi Arabian e-government portal. The official URL of the Saudi Arabian e-government portal is (http://www.saudi.gov.eg). IBM was chosen to be the implementer of the project. Some e-services took a place on the portal through (SADAD). SADAD was established by the Saudi Arabian Monetary Agency (SAMA) to be the national Electronic Bill Presentment and Payment (EBPP) service provider for the Kingdom of Saudi Arabia (KSA). The core mandate for SADAD is to facilitate and streamline bill payment transactions of end consumers through all channels of the Kingdom's Banks. SADAD was launched on October 3rd, 2004 (Yesser, 2013). Customers pay bills through SADAD using available banking channels, and they are: Bank Branch, Automatic Teller Machine (ATM), Online Banking, Phone Banking, and Short Message Service (SMS).

Most government organizations faced many security concerns in the strategy before going online in order to update the citizens' details. Thus, ALELM security Information Company that belongs to the Interior Ministry Company has been established to resolve this issue.

However, the above objectives cannot be achieved without taking into account E-Government Readiness (EGR) on a macro level (i.e., of the overall country) as well as on a micro level pertaining to Organizational E-Government Readiness (OEGR). Evaluating OEGR of public organizations assists in understanding the internal factors linked to each organization that influence OEGR, and that would contribute to the achievement of effective e-government in the public sector. E-government policy makers in Saudi Arabia were concerned about assessing the country's overall EGR since 2006 by applying a qualitative contrast of UN EGR criteria with ICT readiness of Saudi Arabia (Yesser, 2013). Furthermore, Yesser carries out an annual survey to assess OEGR. The survey focuses on the ICT components and skills of the employees within each organization. Yesser is responsible for distributing questionnaires to each organization's minister or general director. This information is collected and serves as a basis for assessing ICT standards and skills of employees in the Saudi Arabian public sector.

The development of Saudi Arabia's e-government program is a continuous process as evidenced by the improvement in the country's ranking in the UN e-Government Readiness Index 2012. The Kingdom of Saudi Arabia is now ranked third among Arab countries and has improved by 17 places to occupy the 41st spot globally in the UN e-Government Readiness Index 2012, demonstrating the country's success in improving its e-services and injecting large-scale investments to increase broadband penetration, which is now expected to reach up to 31% by 2013.

The Ministry of Communication and Information Technology (MCIT), 2013 reports indicate the expectation of much higher demand for online services due to the growing number of Internet users (the number of Internet users grew from around 1 million in 2001 to an estimated 15.2 million by the end of September 9 2012. Internet penetration increased to 52% of the population by the end of 2012. The Fixed Broadband penetration rate stood at around 36.4% of households at the end of 2012), and mobile telecommunications market (the total number of mobile subscriptions grew to around 53.1 million by end of September 2012, with penetration standing at 181.2%. Prepaid subscriptions constitute the majority (85%) of all mobile subscriptions. Total subscriptions to mobile broadband reached 11.73 million by the end of September 9 2012, representing a penetration percentage of 40% of the population (MCIT, 2013).

E-READINESS ASSESSMENT MODELS

The significance to public sector assessment of its readiness relies on evaluating its distinctive opportunities and challenges. For developing countries, an e-readiness assessment can help establish basic benchmarks for regional comparison by market verticals and for national planning. Although many existing e-readiness assessment models show a difference regarding objectives, methodologies and results, there is no comprehensive assessment model. In general, e-readiness assessment models cover one or more of the following factors: ICT infrastructure (the telecommunications infrastructure, Internet access, bandwidth, pricing, and reliability), ICT use levels throughout society including (homes, businesses, schools, and government), Human capacity (literacy, ICT skill levels, and job training), Policy environment (the legal and regulatory environment affecting ICT sector and ICT use(telecommunications policy, trade policy, e-commerce taxation, universal service provisions, consumer protection, and privacy); and ICT economy.

Many e-readiness assessment models have been developed to measure a country or economy's e-readiness over the past two decades. APEC (2000 & 2008) developed an e-commerce readiness assessment guide. United Nations Economic Commission for Europe (2008) proposed diverse indicators to assess e-readiness. These indicators are interdependent and fall within eleven groups including: network access, networked society, network policy, media, networked economy, networked learning, intellectual capital, and labour force.

Bakry (2004) developed a framework for e-readiness assessment model (STOPE) consisting of five categories as follows: Strategy, Technology, Organization, People, and Environment. However, Peters (2005) further examined the criteria used in assessing s country's e-readiness as follows: legal and regulatory environment for ICT use, appropriateness of ICT, affordability of ICT in the local context, ICT capacity and training, availability of locally relevant content and services, use of ICT in business, integration of ICT into peoples' lives, physical access to ICT, socio-cultural factors that affect ICT use, security and peoples' trust in ICT, macroeconomic environment affecting ICT use, and the government's role in driving e-readiness.

Azab, Kamel, and Dafoulas (2009) developed a framework for e-readiness assessment consisting of strategy, processes, technology, and people but do not provide conclusive results for the impact of strategy on e-government e-readiness.

The Center for International Development and Harvard University (2010) proposed an approach for e-government readiness assessment that allows systematic adaptation to the local needs in developing countries and its application to different levels of government. This approach was derived from their model for e-Readiness assessment including network access, networking learning, networked society, networked economy, and network policy.

Despite the varied coverage by different models, the analysis of the literature reflects a consensus on the role of factors such as ICT infrastructure, e-service offered to the users (users can be citizens, other government agencies, businesses, and employees), environment, people skills, accessibility, and economic factors (Alghamdi et al., 2011a; Alghamdi et al., 2011b). Therefore, in this chapter, we focus on these three integrated factors: E-Service offered to the users, ICT infrastructure, and Human skills.

E-READINESS ASSESSMENT METHODOLOGY

There are many models and tools used to assess e-readiness. Each assessment tool or model has a different underlying goal and definition of e-readiness. Therefore, no model will fit each user's requirements. The user should choose carefully and with a clear understanding of the focus of each model.

Research Variables

This chapter measures e-government readiness in the public sector by the status and progress on three integrated factors. These factors were derived from a study done by Harvard CID (2002) “Readiness for the Networked World: A Guide for Developing Countries” which extended a previous APEC study “E-Commerce Readiness Assessment Guide” (APEC, 2000). In addition, the e-services factor derived from the Yesser research team which considered e-services indicated how far the organization is ready for e-government. ICT infrastructure is recognized to be one of the main challenges for e-government in the Kingdom of Saudi Arabia (Alshehri & Drew, 2010; Yesser, 2005). Additionally, human capacity is another major challenge given the lack of ICT skills in the public sector. This is a particular problem in Kingdom of Saudi Arabia (Alshehri & Drew, 2010).

Research Tool

In this work, a questionnaire was designed to investigate the participant's perception towards the ICT environment in public organizations in the Kingdom of Saudi Arabia. The questionnaire had 10 items to assess e-services offered to users, 10 items to assess ICT infrastructure and 10 items to assess human skills as detailed in the discussion section. The participants were asked to indicate the extent of their agreement or disagreement on a seven-point Likert-type scale, ranging from 1 (strongly disagree) to 7 (strongly agree) (Churchill, 1979).

Sample Size

The survey was conducted in the Kingdom of Saudi Arabia over a three-month period (January, February, and March) in 2012. The sample was chosen from government organizations in Saudi Arabia with potential for integrating e-government services. Surveys were distributed to 378 participants from fifty (50) different Saudi Arabia government sector segments as illustrated in Figure 2.1. Nine respondents were from the Ministry of Commerce and Industry, six respondents from each of the following: Hail University, Ministry of Justice, Almadinah Islamic University, Ministry of Health and General Auditing Bureau, and the rest of 44 organizations falling in the range 1 to 5 respondents.

Collected Data

One hundred and fifty six responses were received (a 41% response rate). The respondents were classified according job title and experiences as Figure 1 illustrates. Respondents were from 12 job positions including IT managers (22%), Electronic Services Dep. Managers (13%), IT Development &

Application Managers (12%), IT Consultants (10%), Steering Committee of E-government Managers (9%), System Designers& Analysts (7%), Computer Programmer (7%), IT Project Managers (6%), Network Dep. Managers (5%), IT Help Desk Managers (4%), IT Planning Managers (4%) and CIOs (1%).

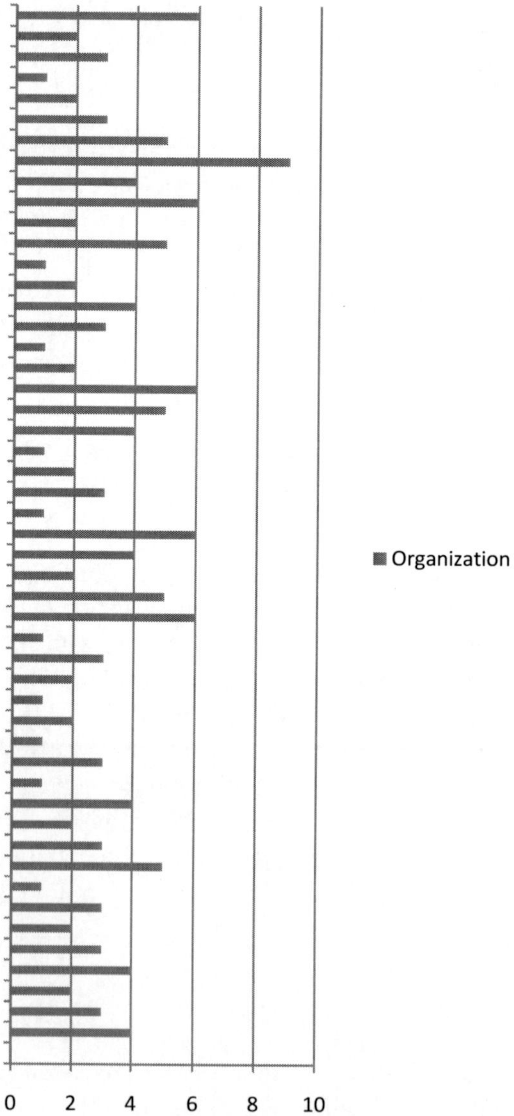

Figure 1. Government Organizations Involved in this Survey.

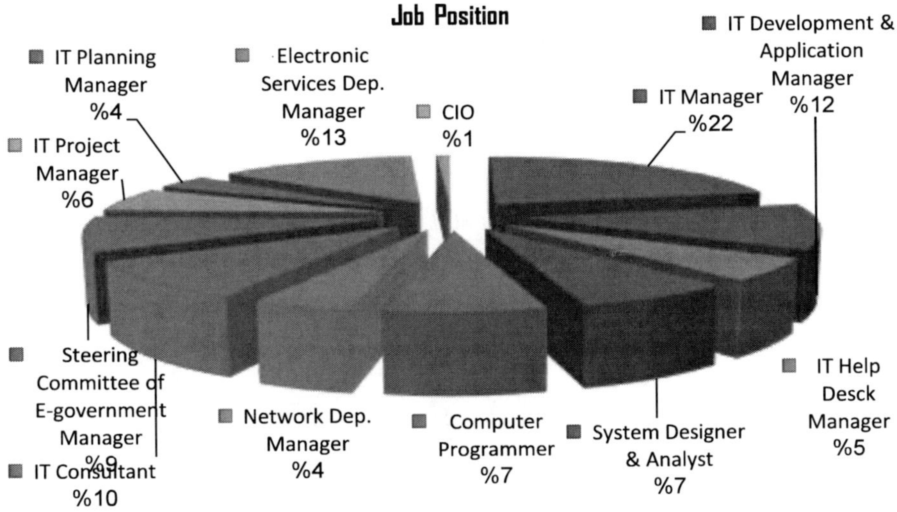

Figure 2. Survey participants: Distribution according to job position.

Assessment Results

The findings indicate the level of readiness among government organizations in the Kingdom of Saudi Arabia as detailed in the in the following section.

Discussion

E-Services

Figure 3 shows the number of services (manual and automatic) offered by organizations. A large number (75%) of organizations offer less that 50 services, but rates tend to decrease when we move from 50-100, 101-200 and above 201 at 8 per cent, 10 per cent and 4 per cent respectively.

Figures 4, 5, 6 and 7 show the percentage of e-services in government organizations in terms of G2C, G2G, G2B and G2E services. Most organizations offer less than 10 e-services. However, there is also a high percentage of government organizations that do not offer any G2E, G2B, G2G and G2C services at 33, 31, 16 and 12 percentages respectively.

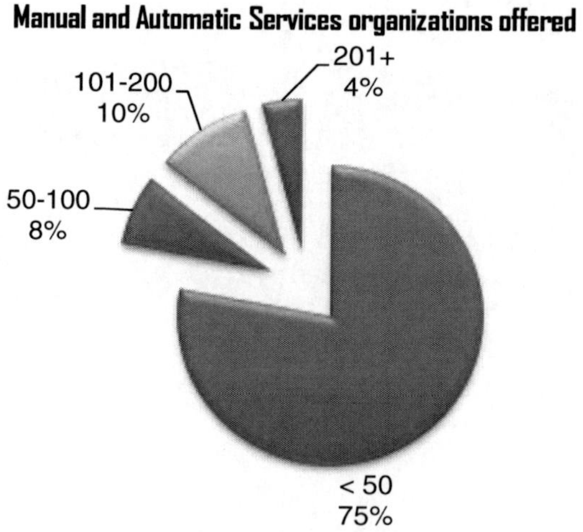

Figure 3.Number of e-services offered by organizations.

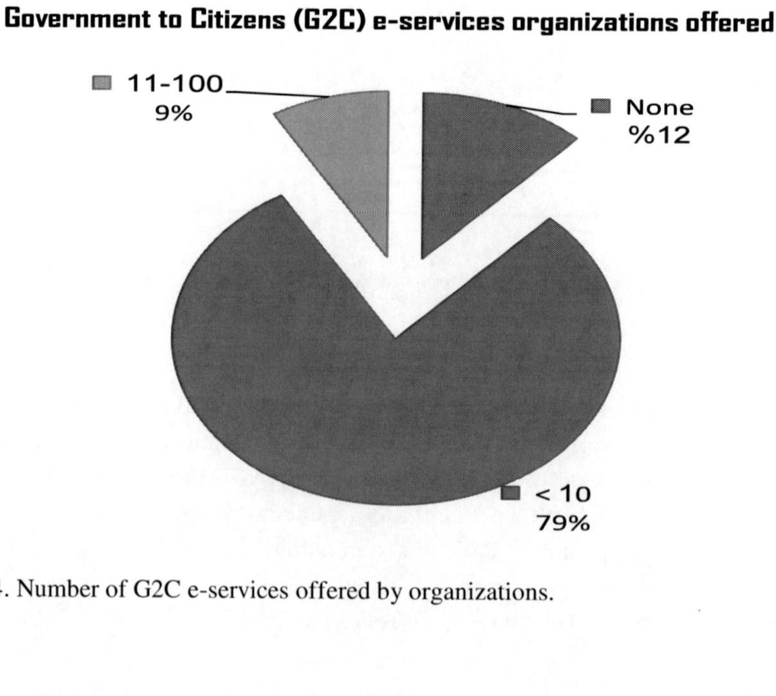

Figure 4. Number of G2C e-services offered by organizations.

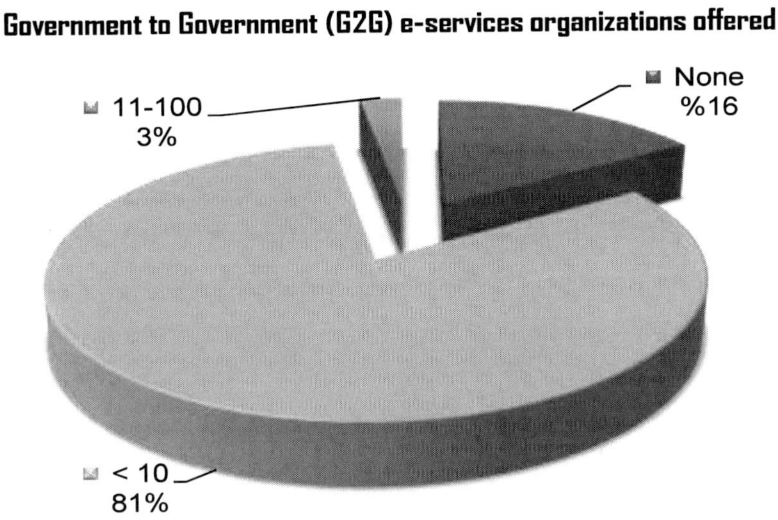

Figure 5. Number of G2G e-services offered by organizations.

Figures 8, 9 and 10 show the percentage of services in government organizations from fully, partially, and non-automated services.

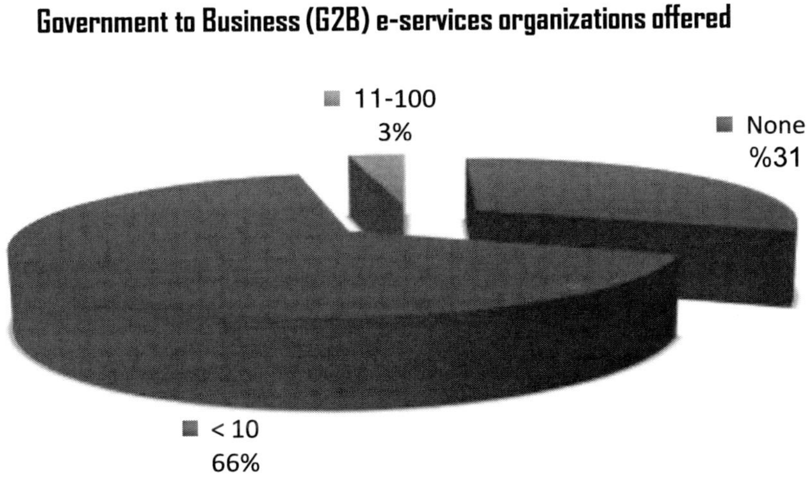

Figure 6. Number of G2B e-services offered by organizations.

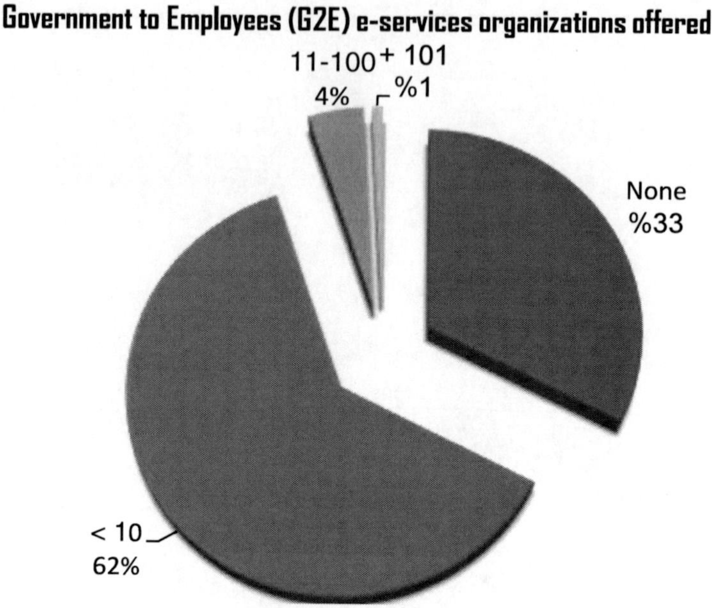

Figure 7. Number of G2E e-services offered by organizations.

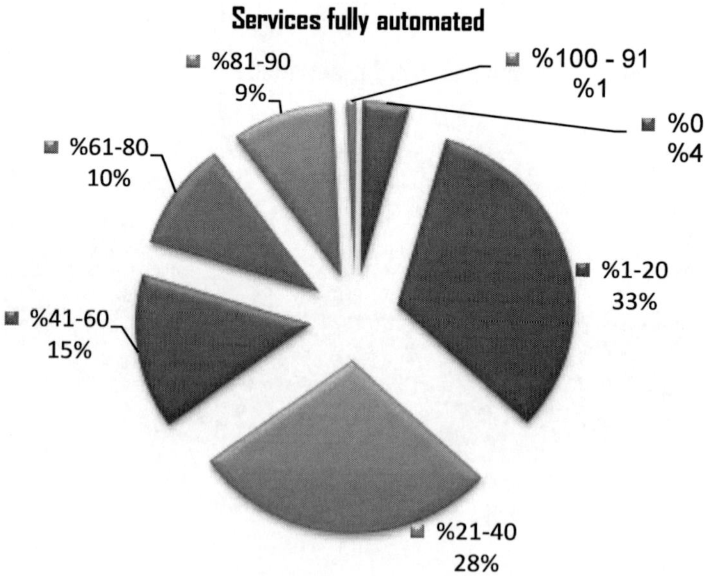

Figure 8. Percentage of fully automated services offered by organizations.

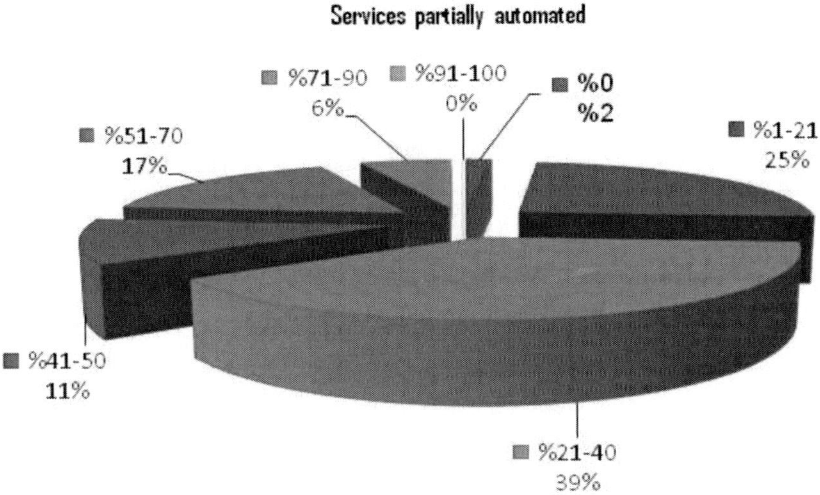

Figure 9. Percentage of partially automated services offered by organzations.

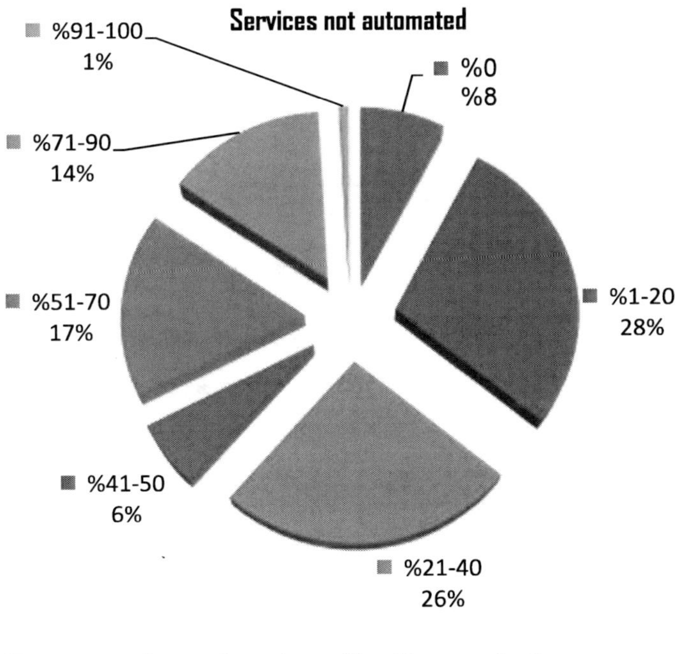

Figure 10. Percentage of manual services offered by organizations.

ICT Infrastructure

When investigating network coverage, results show that around 81% of government organizations have over 80 desktops per 100 employees and the coverage of LAN and WAN is high at 81 and 78% respectively. The ease of access ICT communication network is also high at 80%. Between 61 and 69 % of employees have access to Internet and email; whereas the employee's access to PCs through the home (> 5%) as opposed to a relatively low percentage (around 45%).The percentage of users with high-speed internet (>4Mbps) is between 60 to 74 % (see Table 1).

Table 1. Percentages of desktop and network coverage, email and internet usage, and connection speed

Item	Percentage
Desktop users within the organization (> 80%).	81
The coverage of Local Area Network (LAN) over the organization (> 80%).	81
The coverage of Wide Area Network (WAN) over the organization's branches and customers (> 80%).	78
The ease of access ICT communication network.	80
The employees use official email internally in all departments and sections.	69
The employees use official email externally with all the government and the private sectors.	61
Internet users within the organization (> 80%).	62
The employee's access to PCs through the home (> 5%).	45
The connection speed supported by organization's infrastructure between (385kbps-.5Mbps).	74
Users who have dedicated or other high-speed >4Mbps digital access to the Internet (> 50%).	60
Average	69.1

Human Skills

Regarding IT skills, the technical experience, expertise, IT jobs and qualified IT employees and end user computer skills in government organizations only fall between 44 to 52% which is below average. ICT security training and methods, tools or training provided to assist individuals who lack ICT skills are low at 45 and 49% respectively; whereas training and support keep personnel ICT skill sets current with e-government developments within average at 69% in contrast with the low average of training in advancing employee abilities to involve in e-services. (see Table 2)

Table 2. Percentages of employee's skills and training

Human Skills	Percentage
ICT/IT director/manager has bachelor or above degree in a computer field.	69
ICT personnel held accountable for achieving the organization's e-government goals.	49
The efficiency of jobs that require IT skills.	44
The efficiency of enough qualified IT staff.	46
Incentives provided that attract and retain qualified, highly ICT skilled personnel and encourage superior performance.	50
The technical experience and expertise of e-government teams (internal or partner).	47
The proportion of employees who have International Computer Drive License (ICDL) or/and Internet and Computing Core Certification (IC3) to deal with IT systems.	52
On-going ICT security training is conducted for all relevant personnel.	45
Methods, tools or training is provided to assist individuals who lack the ICT skills or have special requirements for accessing electronic services.	49
Training and support keep personnel ICT skill sets current with e-government developments.	69
Average	52

CONCLUSION

Despite efforts by public organizations in the Kingdom of Saudi Arabia to provide e-services and to support e-government given the political agenda, organizations are not adequately ready for transforming to e-government. The overall level of e-services offered by the public sector is low. Nevertheless, the level of infrastructure is promising as 69.1% of the participants in these organizations agreed that the ICT infrastructures that connect their organizations to the local and public network is ready for applying e-government and they are satisfied about the performance of their organization towards applying e-government. However, only 52 per cent of the participants revealed that employee IT skills are ready to apply e-government. Finally, there are several trends in the government sector in the Kingdom of Saudi Arabia towards the expansion of e-government, which will require the following: i) improving co-ordination and compatibility with Yesser, ii) improving co-ordination and collaboration among organizations, iii) increasing number of e-services provided, v) enhancing ICT infrastructure and connectivity speed internal and with other public organizations, and iv) paying more attention to the attraction, development and retention of talented staff by appropriate remuneration and specialized e-government training.

REFERENCES

Alghamdi, I., Goodwin, R. and Rampersad, G. (2011a) E-Government Readiness Assessment for Government Organizations in Developing Countries, *Computer and Information Science*, 4(3), 1913-1938.

Alghamdi, I., Goodwin, R. and Rampersad, G.(2011b) A Suggested E-Government Framework for Assessing Organizational E-Readiness in Developing Countries. *Informatics Engineering and Information Science*. 479-498.

Alshehri, M., & Drew, S. (2010). Challenges of e-Government Services Adoption in Saudi Arabia from an e-Ready Citizen Perspective. *World Academy of Science, Engineering and Technology, 66*.

APEC. (2000). e-Readiness Assessment Guid Retrieved 15 Sep., 2010, from http://www.apdip.net/documents/evaluation/e-readiness/geosinc01042002.pdf

APEC. (2008). e-Commerce Readiness Assessment Initiative: APEC.

Azab, N. A., Kamel, S., & Dafoulas, G. (2009). A Suggested Framework for Assessing Electronic Government Readiness in Egypt. *Electronic Journal of e-Government* 7(1), 11 – 28.

Bakry, S. H. (2004). Development of e-Government: A STOPE view. *International Journal of Network Management 14*(4), 339-350.

Center for International Development & Harvard University. (2010). Global information technology report; readiness for the networked world Retrieved 28 Sep 2010, 2010, from http://www.weforum.org/gitr.

Churchill, G. A. 1979. A Paradigm for Developing Better Measures of Marketing Constructs. *Journal of Marketing Research,* 16**,** 64-73.

InfoDev (Ed.). (2002). *The E-Government Handbook for Developing Countries*. Washington, D.C.: A project of InfoDev and the Centre for Democracy and Technology.

MCIT. (2013). *ICT indicators in K.S.A (Q3-2012).* Riyadh: Ministry of Communications and Information Technology Retrieved from http://www.mcit.gov.sa/english/Development/SectorIndices/.

O'Neill, R. (2009). *E-government: Transformation of Public Governanace in New Zealand?*, Victoria University of Wellington, New Zealand

OECD. (2003). *e-Government Studies: The E-Government Imperative*: OECD Publishing.

Patel, F., Sooknanan, P., Rampersad, G., and Mundkur, A. (2012) *Information Technology, Development, and Social Change*, New York: Routledge. ISBN 13: 9780415502689.

Peters, T. (2005). E-Ready for What? E-Readiness in Developing Countries: Current Status and Prospects toward the Millennium Development Goals *infoDev, bridges.org.*

United Nations. (2008). UN E-Government Survey 2008 From E-Government to Connected Governance, Department of Economic and Social Affairs *Administration and Development Management, United Nations Online Network in Public Administration and Finance (UNPAN).* USA: Division for Public, United Nations Publications.

Yesser. (2005). Supporting plans and methodologies: (a) E-Readiness assessment methodology for government organizations Retrieved 3, 2013, from http://www.yesser.gov.sa/en/Pages/default.aspx.

Yesser. (2013). E-government Transactions' Methodologies & Handbooks Retrieved 3, 2013, from http://www.yesser.gov.sa/en/Methodologies/mechanisms/Pages/e_government_transactions.aspx.

In: Technology Innovation Leadership … ISBN: 978-1-63117-388-2
Editors: G. Rampersad and F. Patel

Chapter 3

IMPROVING e-GOVERNMENT PERFORMANCE THROUGH ENTERPRISE ARCHITECTURE

M. Ali Hanafiah ***** ***and Robert Goodwin***
Flinders University, Australia

ABSTRACT

Enterprise Architecture (EA) is recognized as a tool that has a holistic approach in the Information Systems (IS) discipline. EA envisages the alignment of technology and business. EA has gained attention not only from private enterprises but also governments. Although interest in adopting EA by government is growing, research that shows how EA can foster e-Government performance is still in its infancy.

Thus it is important to investigate the relationship between the two. By looking at the benefits that EA can generate and variables to determine e-Government performance, this chapter proposes a model to delineate the correlation between EA and e-Government performance.

* Corresponding author: ali.hanafiah@flinders.edu.au

INTRODUCTION

Government transformation through e-Government has become one of the most interesting issues in Information and Communication Technology (ICT) related programs (Wu, 2007). However, Wu (2007) argues that the transformation process has overlooked the importance of interoperability between business and IT. Therefore the boundary between government institutions is yet to disappear. A holistic approach to solve the interoperability issue is needed, as it will clarify in which direction governments should better spend their resources. As a result e-government performance will be improved.

The performance of e-Government is in line with its maturity (Peters, Janssen, & Engers, 2004). Academia, private consultants and non-profit organizations have proposed ways to measure its maturity. Peters et al. (2004) identified three groups of instruments to measure e-Government: stage models, service quality, and performance indicators. In the stage model grouping, there are some models that are already available to use, such as: a four stage model (Layne & Lee, 2001), Siau and Long (2005) proposed with a new model after synthesizing several models. Waseda University in their last international e-Government ranking, have included EA as one of the parameters used in their model (Obi, 2012).

EA is believed to have the capability to incorporate the needs of interoperability (Chen & Daclin, 2006, 2007; Daclin, Chen, & Vallespir, 2006; Hjort-Madsen, 2006) and alignment of business and technology (Gregor, Hart, & Martin, 2007; Martin, 2005; Pereira & Sousa, 2005; Seppanen, Heikkila, & Liimatainen, 2009; Shah & Kourdi, 2007).

It focuses not only on the technological point of view but also non-technological ones (Finkelstein, 2006; Zachman, 1997; Zachman & Sowa, 1992). The capability of EA has attracted the attention of the governments of developed countries and they have adopted it to enhance their e-government initiatives (AGIMO, 2007, 2009; CIO Council, 1999; Hanafiah & Goodwin, 2011; Liimatainen, Hoffmann, & Heikkilä, 2007; Pheng & Boon, 2007; Saha, 2009, 2010).

Unfortunately examples of the use of EA in developing countries e-government programs are still hard to find (Hanafiah & Goodwin, 2011). Developing countries have encountered significant problems in enabling e-Government initiatives (Dada, 2006; Stanforth, 2006).

Despite benefits offered, the development and deployment of EA is not easy and therefore costly. In some cases, public sectors and private companies did not get the expected benefits (Sessions, 2006). In order to find out if EA has any impact on the development of e-Government, it is important to understand the relationship between EA and e-Government. Governments need to be aware of the potential benefits and problems that EA may present. This research seeks to develop a model that can be used to answer the above mentioned question.

To do that, this chapter is constructed as follows: first, some relevant articles are reviewed to find out how to determine and measure factors that will contribute to e-government performance; second, values or benefits that EA may deliver are reviewed; third, the relationship between the influential factors and e-government performance is examined; and lastly, a model is proposed to show how EA can influence e-Government performance.

Measuring e-Government Performance

Governments in developed and developing countries have invested significant resources that include money, time and people with different expertise to develop and deploy e-Government. E-Government is defined as 'the use of ICTs ... as a tool to achieve better government' (OECD, 2003). Therefore through e-Government, ICT will enable: improved accessibility to government information and services, escalate the deliverables, and advance stakeholders interaction with government (Löfstedt, 2005).

Given that definition, Dada (2006) argues that e-Government is not only about computerizing government systems but also about the role of technology in enabling different kinds of improvement. Thus, instead of replacing how government delivers the information, ICT is adding on value to service delivery (Coursey & Norris, 2008).

In applying e-Government, Heeks and Davies (1999) estimated around US$500 billion per year is spent on ICT related systems in the public sector. However, the value of this investment is often hard to measure. One way to quantify the value of e-Government performance is by measuring the maturity level of e-Government initiatives.

Having understood the maturity level of the initiatives, governments will be able to advance and retain their competitiveness (De Bruin, Freeze, Kaulkarni, & Rosemann, 2005).

As a result they will know how to improve the quality and yet control the effectiveness. Current models that are available can be assembled into stage models, services quality and performance indicators (Peters et al., 2004).

Table 3.1. E-Government Stages

Author	Stage 1	Stage 2	Stage 3	Stage 4	Stage 5	Stage 6
Hiller and Bélanger (2001)		Information dissemination	Two-way communication	Integration	Transaction	Participation
Layne and Lee (2001)		Catalogue	Transaction	Vertical Integration	Horizontal Integration	
Moon (2002)		Information dissemination	Two-way communication	Service and financial transaction	Vertical and Horizontal Integration	Political participation
Shahkooh et al. (2008)		Online presence	Interaction	Transaction	Transformation	Digital Democracy
Siau and Long (2005)		Web presence	Interaction	Transaction	Transformation	e-Democracy
The United Nations (2003)		Emerging presence	Enhanced presence	Interactive presence	Transactional presence	Network presence
Wescott (2001)	e-mail and internal network	Internal organization & public access to information	Two-way communication	Exchange of value	Digital democracy	Joined-up government

Stage Models

Stage models identify the maturity of e-Government through how governments exploit ICT, such as the internet, to interact with their stakeholders e.g.: citizens, private sectors, and other government bodies. Although most of the models in this group start with the static site of a government website, the paths by which each of the models progresses is different. Most of the models start with the official presence of an e-government website characterized by simple information distribution (Hiller & Bélanger, 2001; Layne & Lee, 2001; Moon, 2002; Shahkooh, Saghafi, & Abdollahi, 2008; Siau & Long, 2005; Wescott, 2001). Table 3.1 shows the e-government stage evolution from the above mentioned authors in alphabetical order.

Table 3.1 shows that most of the models view e-Government from how governments utilize it to improve public participation (Bellamy & Taylor, 1994; Danziger & Andersen, 2002; De Araujo, 2001; Edelenbos & Klijn, 2006; Homburg, 2008; Torres, Pina, & Royo, 2005). Therefore those models can be used to measure e-Government where the Internet is used extensively. These models are however lacking in empirical evidence. For that reason Coursey and Norris (2008) conducted an empirical study to find out whether the above mentioned models can be used to ensure that e-government is progressing as predicted in the model.

The United Nations has been adopting the stage model to measure worldwide e-Government readiness (The United Nations, 2003, 2004, 2005, 2008, 2010, 2012). Research by Coursey and Norris (2008) found that it cannot be clearly perceived that e-Government is progressing based on the proposed stages. The stage model may still be relevant to recognize at what stage the e-Government is, but relying solely on one of the models may lead to insufficient interpretation of e-Government performance because the quality of public service delivery is excluded in the model.

Service Quality

The main role of a government is delivering services to the people (De Araujo, 2001; Kellough & Nigro, 2006). The characteristic of service is intangible (Parasuraman, Zeithaml, & Berry, 1985), it requires an instrument that can measure the effectiveness of how a service is being delivered, e.g., service quality. Service quality relies on satisfaction and perceived quality

(Peters et al., 2004). It was Parasuraman, Zeithaml, and Berry (1988) who designed an instrument called SERVQUAL to measure perceptions of service quality. Parasuraman et al. (1988) argue that customers' expectations and perception of quality can be used to measure service quality. In order to do so, they developed a 22-item questionnaire in five dimensions, namely: 'tangibles, reliability, responsiveness, assurance, and empathy' (Parasuraman, Berry, & Zeithaml, 1991; Parasuraman et al., 1988).

Parasuraman et al. (1988) categorize physical facilities, equipment and personnel as part of the tangibles dimension. They define the reliability dimension as accuracy and the ability to deliver the promised service. Willingness to give immediate support is put into the responsiveness dimension. As for the assurance dimension, they include sufficient knowledge to stimulate trust and confidence. In the last dimension, empathy, understanding individual customer profiles is used.

Initially, SERVQUAL was proposed in marketing studies. Then some Information Systems (IS) studies adopted the model to measure IS effectiveness (Gorla, 2012; Kaisara & Pather, 2011; Kettinger & Lee, 1994; Myerscough, 2002; Petter, DeLone, & McLean, 2008; Pitt, Watson, & Kavan, 1995; Van Dyke, Kappelman, & Prybutok, 1997; Watson, Pitt, & Kavan, 1998). Although the five dimensions in SERVQUAL have been stable over time, these dimensions cannot be considered universal for all disciplines (Bigne, Moliner, & Sanchez, 2003). Therefore the use of SERVQUAL in IS research requires modification (Bigne et al., 2003; Gorla, 2012). Gorla (2012) modified the instrument to measure IS success. This modified instrument has become prominent because it can measure comprehensive ICT service deliverables from an organization rather than individual products. Thus it is able to measure whether or not the expected benefits have been achieved.

Performance Indicators

In determining the performance of government web sites, Lee (2003) proposed some measurements based on modifications to the Simeon (1999) studies. His proposed measurement components are as follow:

1. *Attracting*. It is meant to impress the users and is determined by several items i.e.: the logo and tagline; graphics this includes the choice of colors, layout, themes; self-advertisement from the web

site's owner; attracting services such as quizzes, maps; features for attracting such as a gallery, tourism, game, finance.

2. *Informing*. It consists of modified Simeon (1999) components i.e.: local links, content for publicity, content for learning, reports, descriptions of the institution, description of the online administrative services, projects, contact information and counseling.
3. *Community* comprised of ten items: online forum, events, partner links, newsletter, message boards, users' participation, focus of news, vision or values, domain identity and community services.
4. *Delivering* is a binary variable to identify the presence or absence of features for each essential item such as a search engine, mailing list, framework, multimedia, password system, downloadable publications.
5. *Innovation* is provided to measure whether or not public institutions have utilized the internet with sufficient service innovation.

Although Peters et al. (2004) put this two-stage model for monitoring website strategy in the performance indicators group, this model only measures what is visible on the interface of a web site. This group is viable in assessing what governments offer on the internet web site. Yet this model may not be suitable to find out how government is performing in delivering their services based on the utilizations of ICT that needs to be aligned with how governments conduct their businesses.

The Benefits of Enterprise Architecture

A sound alignment between ICT and Business will definitely deliver benefits to any organization that is able to put it into action (Gregor et al., 2007; Pereira & Sousa, 2005). As a result the outcomes will be improved. Strategic alignment theory defines alignment as a never ending and continuous improvement through understandable performance indicators, enterprise framework, good governance and harmonious execution processes (Venkatraman, Henderson, & Oldach, 1993). In constructing this alignment, organizations have made ICT their main tool to deliver business objectives (Gregor et al., 2007). The fundamental alignment should be clearly defined in their business and ICT strategic planning.

In order to achieve an alignment of ICT and business that includes resources such as human capital, organization and knowledge, it is important to have comprehensive management and governance instruments. One tool that can be used to generate this alignment is enterprise architecture. Currently, there is no single definition of Enterprise Architecture. Langenberg and Wegmann (2004) define EA as 'a blueprint that documents all the information systems within the enterprise, their relationships, and how they interact to fulfill the enterprise mission'. Enterprise architecture will play an important role in achieving a visible integration of business and ICT in a common framework (Gregor et al., 2007). Thus, it is important for enterprise architecture to contain detailed information on the expected business and ICT processes, including the required strategies and resources.

As an emerging topic in the field of information systems, EA has attracted interest from both the private and public sectors. Zachman (1987) introduced a framework for Enterprise Architecture and enriched it in 1992 (Zachman & Sowa, 1992). Many efforts in the area of EA frameworks and methodologies have been presented. The Open Group developed the Open Group Architecture Framework (TOGAF). It covers not only the framework but also the methodology in a holistic enterprise view (The Open Group, 2009). In 1999, the Chief Information Officers Council successfully formulated the Federal Enterprise Architecture Framework (FEAF) (CIO Council, 1999). The Singaporean Government formulated what is called the Singapore Government Enterprise Architecture (SGEA) (Pheng & Boon, 2007; Saha, 2009).

Measuring EA Effectiveness

It is important to have an instrument to measure the effectiveness of EA. Thus the quality of such initiatives can be measured and improved (De Bruin et al., 2005). Luftman (2003) argues that it is vital to have a sufficient maturity model to achieve a proper implementation strategy. In the field of EA, Hanafiah and Goodwin (2012) reviewed some EA maturity models. The comparison of those models can be seen on Table 3.2.

Hanafiah and Goodwin (2012) discovered that each model has apparent strengths and weaknesses. Ross, Weill, and Robertson (2006) presented a concise model in which the alignment of business and ICT has already in place. They presented relevant examples to show how this model is used. Therefore the model is relatively easy to follow and links with real well-known industries such as insurance, retail, and health. Because this model mainly focuses on an industry which has already started EA initiatives, new EA entrants will not find this suitable for their interests. An early adopter will not find it useful for working out how to put EA in place.

Table 3.2. EA Maturity models comparison

Maturity model	Level completeness	Shifting strategy	Measurement Details	Usability	Diagnostic tools
Enterprise Architecture Maturity Model (NASCIO, 2003)	√	Not Given	Sufficient	General	√
Gartner's IT Score (Burke, 2010)	√	Not Given	Sufficient	General	⊗
GAO's EAMMF (Hite, 2003)	√	Brief	Sufficient	Specific	√
Four stages architecture maturity (Ross et al., 2006)	⊗	Available	Fair	General	⊗
The IFEAD's E2AMM v2 (IFEAD, 2004)	√	Not Given	Rich	General	⊗
OMB's EAAF (Office of Management and Budget, 2005)	√	Not Given	Sufficient	Specific	√

Despite rigid measurement elements, E2AMM v2 does not provide any diagnostic tools that will help in using this model. Gartner's model and NASCIO's model (2003) are concise and relatively easy to follow. As they were developed only to measure the maturity level, insufficient attention is paid to the other criteria used in this research.

Both the US GAO and the OMB focus on measuring US government EA initiatives. To date, evidence could not be found showing that this model has been used in other industries.

In light of the above, using the meta-synthesis approach (Noblit & Hare, 1988), Hanafiah and Goodwin (2012) proposed an alternative maturity model. This model is built upon simple yet comprehensive principles. This model is summarized in Table 3.3. The model is called the six As EA maturity model. The benefits of this model are:

1. Refining the existing models which are not fully comprehensive from criteria used in their research.
2. Taking early adopters of EA into consideration. Current models are not entirely useful for various EA adopters.
3. Providing a perceptible shifting strategy from one level to another.

Table 3.3. The six As EA maturity model

Lvl	Level Name	Indicator
1	Absence	EA does not exist
2	Awareness	Awareness is built, high ranking managerial approval is gained
3	Acceptance	EA organization is established, involvement of more internal parties
4	Application	EA is used as reference in aligning Business and ICT
5	Alignment	Continuous alignment of Business/ICT is regularly conducted
6	Augmentation	EA value is extended to indirect stakeholders

It is likely that any enterprise interested in developing and deploying an EA will be able to utilize this model. A brief explanation of each level in this model is as follows:

Absence. At this level, the organization does not recognize the need for an EA despite the level of ICT usage in the business. Typically, in managing change, the organization would use the trial and error approach. To shift from this level, it is important to have someone either internal to the organization or an external consultant influence high-level officials by showing the importance of EA. This achievement is important because EA will be more effective if applied in a top-down fashion (Finkelstein, 2006).

Awareness. Following the acceptance of EA by top-level officials, awareness of EA should then be extended to relevant parties. An initial ad-hoc committee can be used to start formulating or researching the EA framework and methodology to be used.

Acceptance. At this level a formal unit that is responsible for developing and maintaining EA has been established. It will include the involvement of other stakeholders who will be directly affected by the EA. A formal model of EA should also be available at this stage. The alignment of business and ICT has not been established however.

Application. One feasible indicator at this level is that Business and ICT transformations refer to the EA. EA becomes the main resource in fulfilling the needs of the core system. This core system is represented by digitizing core data or business processes within the organization (Ross et al., 2006). As enabling EA is a continuous process (Gregor et al., 2007), regular review of the architecture will be needed to satisfy supporting systems.

Alignment. Internal systems, core and supporting, to run the business are developed with the guidance of EA. Business organization will also then be adjusted. In order to move to the next level, it is essential to create a channel by which external stakeholders, who are indirectly affected by the systems, can easily input their concerns or views. This will enhance the systems to fit with their needs.

Augmentation. At the top level of this model, internal stakeholders and indirect parties have gained the benefit of EA enabled systems. They have sufficient access to propose improvements to the systems.

The Enterprise Architecture Benefits Model

Zachman (2003) said that 'enterprise architecture is used for the management of enterprise change'. He further argues that 'if enterprise architecture is not used, there are only three options for managing enterprise change: by trial and error; by reverse engineering; or by going out of businesses'. Consequently the existence of EA is vital because it is not only used to express what the objectives of an enterprise are but also managing changes within.

In the field of EA several researchers mention some benefits of EA such as management of change (Finkelstein, 2006; Zachman, 2003), alignment of business and IT (Aversano, Grasso, & Tortorella, 2012; Gregor et al., 2007; Pereira & Sousa, 2005; Wegmann, Balabko, Lê, Regev, & Rychkova, 2005), cost effectiveness (Finkelstein, 2006; Lankhorst, 2009; Ross et al., 2006; John A. Zachman, 2003). Unfortunately such claims are lacking in empirical evidence (Tamm, Seddon, Shanks, & Reynolds, 2011). For that reason, Tamm et al. (2011) conducted research to find out what values or benefits EA can deliver. After systematic review on extensive articles about the benefits of EA, they proposed an EA Benefits Model (EABM) as can be seen on Figure 3.1.

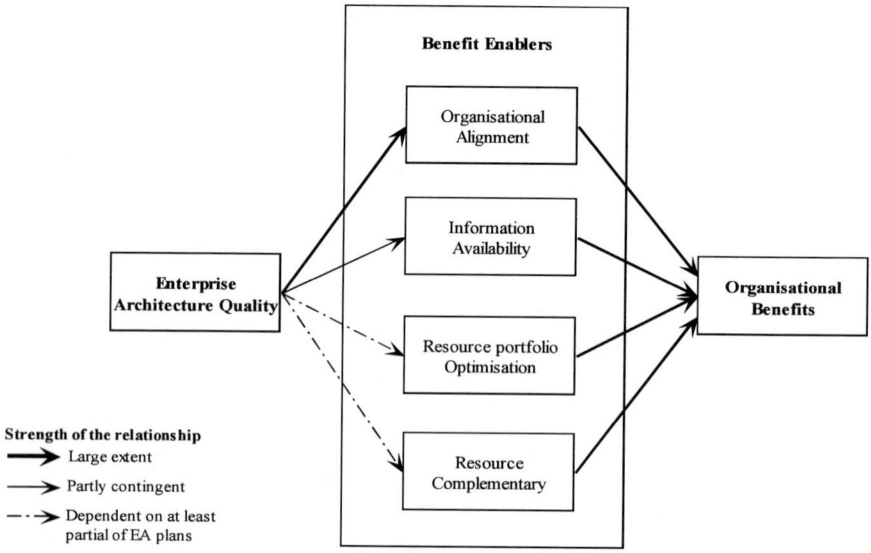

Figure 3.1. EA Benefits Model (EABM).

In the model, Tamm et al. (2011) argue that there is no direct benefit from EA to organization. Instead EA is generating some factors that potentially lead to organizational benefits. Therefore they place the benefit enablers between EA quality and the organization benefits. The dotted lines in Figure 3.1 refer to resource portfolio optimization, and resource complementarity. These variables are dependent on the partial creation of EA. Tamm et al. (2011) further argue that information availability is partly dependent on the existence of EA. As a final point, organizational alignment can be achieved as a result of

EA development activities. For that reason they used a thick line to represent it in the model.

Tamm et al. (2011) define each variable in the EABM as follows:

1. EA Quality is defined where EA has a clear vision of the future of a well-aligned organization. It is the only independent variable in the model. There are two ways for assessing the quality of EA:
   - Directly evaluating the products such as EA documentation.
   - Using EA development process as a proxy.
2. Organizational Benefits is defined as the outcomes that contribute directly to organizational performance.
3. Organizational alignment is the extent to which an organization's subunits share a common understanding of its strategic goals, and contribute towards achieving these goals.
4. Information Availability is the extent of useful, high-quality information accessible to organizational decision makers.
5. Resource portfolio optimization is the extent to which an organization leverages its existing resources, invests in resources that target performance gaps, and minimizes unnecessary investments in duplicated resources.
6. Resource complementarity is the extent to which the organization's resources synergistically support the pursuit of its strategic goals.

In light of the above, the EABM may be used to pinpoint the relationship between EA and e-Government performance.

RELATIONSHIP BETWEEN EA AND e-GOVERNMENT PERFORMANCE

Although there are many studies with regard to the use of EA in government (Janssen & Kuk, 2006; Liimatainen, 2008; Katja Liimatainen et al., 2007; Martin, 2005; Saha, 2010; Seppanen et al., 2009), very few studies have successfully demonstrated the value of EA to Government (Liimatainen, 2008). Tamm et al. (2011) argue that although there is no direct benefit of EA to an organization, the EABM displays the relationship between EA and organizational benefit. The world's largest and considered to be the most complex organization, government, (Saha, 2010) can also adopt the EABM to find out if EA delivers benefit to their organizations. One significant benefit that needs to be perceived is whether or not the e-Government initiative is performing as expected. E-Government performance can then be used as a variable to measure the benefit. Therefore this chapter argues that EABM could be adopted to define the relationship between EA and e-Government performance.

In the EABM, Tamm et al. (2011) mentioned that EA quality can be assessed in two different ways: assessing the final product of EA and using the process of EA development as a proxy. These assessment methods are also used in EA maturity (Banger, 2008; Burke, 2010; Hanafiah & Goodwin, 2012; IFEAD, 2004; NASCIO, 2003) to measure the effectiveness of an EA program within the institution either government or private. Therefore the outcome and the process of EA maturity assessment are similar to EA quality proposed in the EABM. Consequently this chapter proposes EA maturity to measure the quality of EA. Although two variables have been identified and modified to investigate the relation of EA and e-Government, this chapter needs to map the benefit enablers in the EABM and the dependent variables that lead to e-Government performance.

MAPPING EABM'S BENEFITS AND e-GOVERNMENT PERFORMANCE'S VARIABLES

The EABM proposed four variables that lead to organizational benefits. Tamm et al. (2011) identified that it is possible to modify the grouping of the enablers in the model due to the lack of empirical validation. The organizational benefit that this chapter would like to discover is the e-Government performance. Currently, as mentioned in earlier section, various instruments to measure the e-Government performance are already available.

In measuring e-Government performance this chapter will mainly adopt service quality because it can evaluate the overall quality of services (Bigne et al., 2003). Although this instrument was developed in marketing studies (Parasuraman et al., 1991; Parasuraman et al., 1988), some studies in the field of public services (Bigne et al., 2003), information systems (Gorla, 2012; Kettinger & Lee, 1994; Van Dyke et al., 1997), e-Government (Kaisara & Pather, 2011; Peters et al., 2004) are keen to use the model. Therefore this chapter will use modified instruments of SERVQUAL (Parasuraman et al., 1988) in the field of Information System (Gorla, 2012).

To find theoretical relationships, the variance approach is used because those models have different properties of entities (Burton-Jones, McLean, & Monod, 2011). In this manner, it is essential to map the EA benefits and e-Government performance to find out whether or not the EABM's benefits and e-Government performance's variables fit. The mapping of these models is depicted in Figure 3.2.

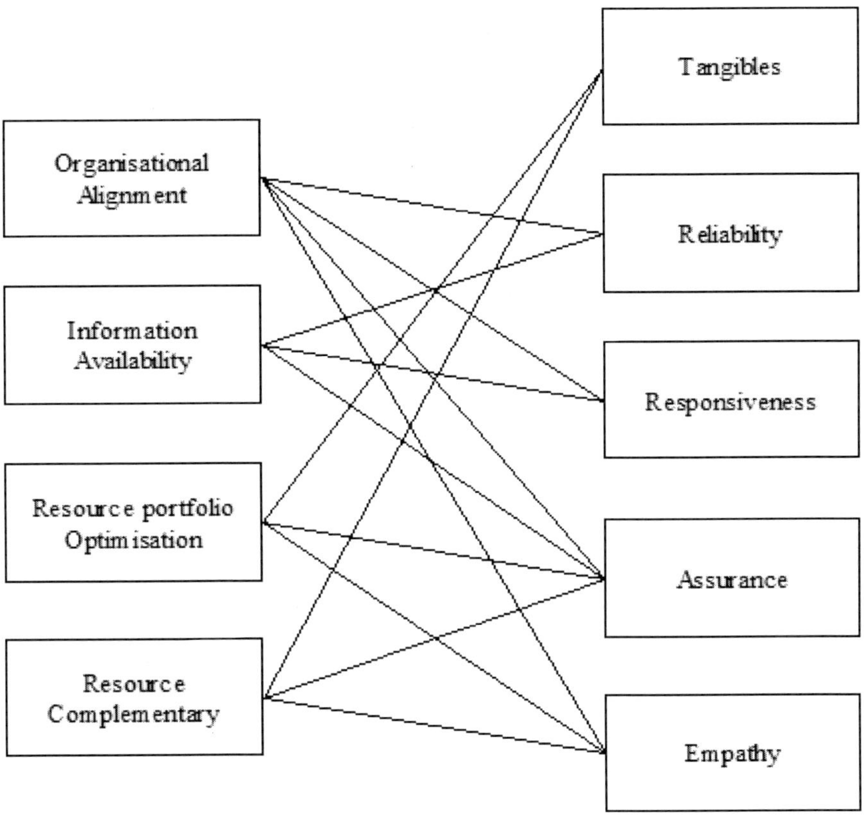

Figure 3.2. Mapping EABM's Benefit Enablers and Service Quality variables.

The EABM defined four dependent variables that may influence the benefit of EA (Tamm et al., 2011), whereas Parasuraman et al. (1988) defines five variables to measure service quality. Organizational alignment in the EABM is used to measure how organizations share a common understanding which in turn improves reliability, responsiveness, assurance and empathy among each other. Information availability is used to find out the usefulness, and accessibility of output from e-Government. Consequently this requires reliable, fully responsive and assured IS. The last two variables i.e. Resource portfolio optimization and resource complementary are used to ensure there will be efficient infrastructure investment. These variables are tangibles. Insufficient infrastructure performance may lead to loss organization's empathy.

Model Development

Having reviewed variables in the EABM and SERVQUAL, it is found that both models are considered to be complementary fit. Therefore this chapter would like to propose the modified EABM to depict the relationship between EA and e-Government performance. Modified instruments from Gorla (2012) for variables in SERVQUAL (Parasuraman et al., 1988) may then be used to measure e-Government performance. The maturity of EA could drive the performance of e-Government through its performance drivers adopted from the EABM. Therefore the more mature EA the better performance that e-Government could deliver. Consequently EA may be used to improve the quality or performance of e-Government. In order to measure this, the following model depicted in Figure 3.3 is used.

The degree of EA maturity will be assessed through completeness of EA documentations and its development process (Hanafiah & Goodwin, 2012; NASCIO, 2003). Although Hanafiah and Goodwin (2012) proposed an alternative EA maturity model, measurement details have not yet been developed. Adopting NASCIO (2003) the following categories are used to measure the maturity level of an EA program: administration to measure governance and responsibilities, planning to gather the road map and execution plan, templates and process used to measure the framework, blueprint is gathered from collection of standards used, communication is assembled from education and distribution of EA, compliance is collected through adherence among EA elements, integration is compiled through management process engagements to exchange information, and involvement to determine user participation in the development and deployment process.

Gorla (2012) modified SERVQUAL instruments to measure the range of facilities used, up-to-date hardware and software, engaging personnel and the operating hours of the IS department for tangibles variable. For reliability, it measures fulfillment of promises, dependability, timely service delivery, service commitment. Prompt responses, willingness, and staff availability were used for responsiveness variable. Some elements are used to measure assurance such as: instilling confidence, feeling safe in using the system, interaction between IS department and users, IS employees' knowledge. Empathy includes individual attention to users, personal approach to users, IS department has user's best interest and understand user's needs. These variables will be used to measure the e-Government performance.

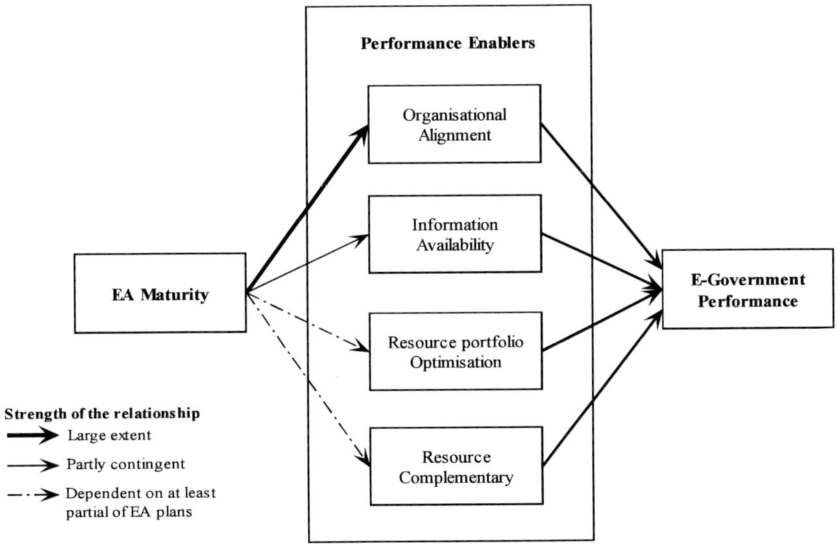

Figure 3.3. Relationship of EA and e-Government Performance.

CONCLUSION

The benefits of EA as they are identified in the EABM have given an opportunity to investigate the relationship between EA and e-Government. Although there is no empirical evidence yet for EABM (Tamm et al., 2011), the benefit enablers remain worthwhile to be included as the performance drivers. By using the variance approach this chapter found out that the enabling factors in the EABM and variables in the SERVQUAL are complementary fit. SERVQUAL has long been established as an instrument that can measure service quality. It has already been empirically tested in the field of IS (Gorla, 2012; Kettinger & Lee, 1994; Myerscough, 2002; Van Dyke et al., 1997). Therefore using SERVQUAL variables for e-Government performance in the model could minimize potential issues in generalizing the findings.

This chapter has just modified the EABM. Further chapter to investigate its use in a government institution is needed. This will ensure that the model is useful. Investigating the model in developing countries will be significant due to their relatively slow improvement in their e-Government readiness (Dada, 2006; The United Nations, 2003, 2004, 2005, 2008, 2012).

REFERENCES

AGIMO. (2007). Australian Government *Architecture Reference Models Version 1.0.*

AGIMO. (2009). Australian Government *Architecture Reference Models Version 2.0.*

Aversano, L., Grasso, C., & Tortorella, M. (2012). A Literature Review of Business/IT Alignment Strategies. *Procedia Technology*, 5(0), 462-474. doi: 10.1016/j.protcy.2012.09.051.

Banger, D. R. (2008). Maturity Assessment for the Enterprise Architecture (EA) Function. Retrieved 31 October, 2011, from http://www.s-ea-t.com/EA-MaturityAssesment.pdf

Bellamy, C., & Taylor, J. A. (1994). Introduction: Expoliting IT in public administration – towards the information polity? *Public Administration*, 72(1), 1-13. doi: 10.1111/j.1467-9299.1994.tb00996.x

Bigne, E., Moliner, M. A., & Sanchez, J. (2003). Perceived quality and satisfaction in multiservice organisations: the case of Spanish public services. *Journal of Services Marketing*, 17(4), 420-442.

Burke, B. (2010). IT Score for Enterprise Architecture. Retrieved 28 October, 2011, from http://my.gartner.com/portal/server.pt/gateway/

Burton-Jones, A., McLean, E. R., & Monod, E. (2011). On approaches to building theories: Process, variance and systems: Working paper, Sauder School of Business, UBC.

Chen, D., & Daclin, N. (2006). Framework for enterprise interoperability. *Interoperability for Enterprise Software and Applications,* 77-88.

Chen, D., & Daclin, N. (2007). Barriers driven methodology for enterprise interoperability. Establishing the foundation of collaborative networks, 453-460.

CIO Council. (1999). Federal enterprise architecture framework. Retrieved 5 July 20011, from http://www.cio.gov/Documents/fedarch1.pdf

Coursey, D., & Norris, D. F. (2008). Models of E-Government: Are They Correct? An Empirical Assessment. *Public Administration Review*, 68(3), 523-536. doi: 10.1111/j.1540-6210.2008.00888.x

Daclin, N., Chen, D., & Vallespir, B. (2006). Enterprise interoperability measurement-Basic concepts. *Enterprise Modeling and Ontologies for Interoperability,* Luxemburg.

Dada, D. (2006). The failure of e-government in developing countries: A literature review. *The Electronic Journal of Information Systems in Developing Countries*, 26.

Danziger, J. N., & Andersen, K. V. (2002). The impacts of information technology on public administration: An analysis of empirical research from the :Golden Age" of tranformation. *International Journal of Public Administration*, 25(5), 591-627.

De Araujo, J. F. F. E. (2001). Improving public service delivery: the crossroads between NPM and traditional bureaucracy. *Public Administration*, 79(4), 915-932.

De Bruin, T., Freeze, R., Kaulkarni, U., & Rosemann, M. (2005). *Understanding the main phases of developing a maturity assessment model.*

Edelenbos, J., & Klijn, E. H. (2006). Managing stakeholder involvement in decision making: a comparative analysis of six interactive processes in the Netherlands. *Journal of Public Administration Research and Theory*, 16(3), 417.

Finkelstein, C. (2006). Enterprise Architecture For Integration : Rapid Delivery Methods and Technologies. London: Artech House.

Gorla, N. (2012, 2012 April-June). Information systems service quality, zone of tolerance, and user satisfaction. *Journal of Organizational and End User Computing,* 24, 50+.

Gregor, S., Hart, D., & Martin, N. (2007). Enterprise architectures: enablers of business strategy and IS/IT alignment in government. Information Technology & People, 20(2), 96-120. doi: 10.1108/09593840710758031

Hanafiah, M. A., & Goodwin, R. (2011). Government Enterprise Architecture: lessons learned from the United States' FEAF and Australia's AGA. *International Information Systems Conference* (ISICO), 1, 97-102.

Hanafiah, M. A., & Goodwin, R. (2012). An Alternative Enterprise Architecture Maturity Model. *IJACP-Proceedings of the International Conference on Business Management & Information Systems*, 509-517.

Heeks, R., & Davies, A. (1999). Different approaches to information age reform. Reinventing Government in the Information Age: International

Practice in IT-Enabled Public Sector Reform. New York: Routledge, 22-48.

Hiller, J. S., & Bélanger, F. (2001). Privacy Strategies for Electronic Government. Retrieved 25 March, 2013, from http://wwwpub.zih.tu-dresden.de/~isih/projekte/keb/literatur/ Privacy_strategies_for_Electronic_Commerce.pdf

Hite, R. C. (2003). A Framework for Asseing and Improving Enterprise Architecture Management version 1.1. Retrieved 31 October, 2011, from www.gao.gov/new.items/d03584g.pdf

Hjort-Madsen, K. (2006, 04-07 Jan. 2006). Enterprise Architecture Implementation and Management: A Case Study on Interoperability. Paper presented at the System Sciences, 2006. HICSS '06. *Proceedings of the 39th Annual Hawaii International Conference on.*

Homburg, V. (2008). Understanding e-government: Information systems in public administration: *Routledge Basingstoke*, UK.

IFEAD, I. F. E. A. D. (2004). The IFEAD Extended Enterprise Architecture Maturity Model (E2AMM). Retrieved 28 October, 2011, from www.enterprise-architecture.info/Images/E2AF/ E2AMMv2.PDF

Janssen, M., & Kuk, G. (2006). A complex adaptive system perspective of enterprise architecture in electronic government.

Kaisara, G., & Pather, S. (2011). The e-Government evaluation challenge: A South African Batho Pele-aligned service quality approach. *Government Information Quarterly,* 28(2), 211-221. doi: http://dx.doi.org/10.1016 /j.giq.2010.07.008

Kellough, J. E., & Nigro, L. G. (2006). Dramatic reform in the public service: At-will employment and the creation of a new public workforce. *Journal of Public Administration Research and Theory,* 16(3), 447.

Kettinger, W. J., & Lee, C. C. (1994). Perceived service quality and user satisfaction with the inf. Decision Sciences, 25(5,6), 737-737.

Langenberg, K., & Wegmann, A. (2004). Enterprise architecture: What aspects is current research targeting. *Laboratory of Systemic Modeling, Lausanne.*

Lankhorst, M. (2009). Enterprise architecture at work: Modelling, communication and analysis: Springer-Verlag New York Inc.

Layne, K., & Lee, J. (2001). Developing fully functional E-government: A four stage model. *Government Information Quarterly*, 18(2), 122-136. doi: 10.1016/s0740-624x(01)00066-1

Lee, J.-K. (2003). A model for monitoring public sector web site strategy. *Internet Research*, 13(4), 259-266.

Liimatainen, K. (2008). Evaluating benefits of government enterprise architecture. *Paper presented at the 31st Information Systems Research Seminar in Scandinavia.*

Liimatainen, K., Hoffmann, M., & Heikkilä, J. (2007). Overview of Enterprise Architecture work in 15 countries. Retrieved 19 October, 2012, from http://www.vm.fi/vm/en/04_publications_and_documents/01_publications/04_public_management/20071102Overvi/FEAR_ENGLANTI_kokonaan.pdf

Löfstedt, U. (2005). E-government-assessment of current research and some proposals for future directions. *International journal of public information systems*, 1(1), 39-52.

Luftman, J. N. (2003). Competing in the information age: align in the sand: Oxford University Press, USA.

Martin, N. J. (2005). Business Strategy and Information Systems Alignment: A Study of the Use of Enterprise Architectures in Australian Government. (Doctor of Philosophy), *The Australian National University*, Canberra.

Moon, M. J. (2002). The Evolution of E-Government among Municipalities: Rhetoric or Reality? *Public Administration Review*, 62(4), 424-433. doi: 10.1111/0033-3352.00196

Myerscough, M. A. (2002). Information systems quality assessment: Replicating Kettinger and Lee's USISF/SERVQUAL combination. Paper presented at the *Proceedings of the Eighth Americas Conference on Information Systems (AMCIS).*

NASCIO. (2003). NASCIO Enterprise Architecture Maturity Model. Retrieved 23 October, 2012, from http://www.nascio.org/publications/documents/NASCIO-EAMM.pdf

Noblit, G. W., & Hare, R. D. (1988). Meta-ethnography: Synthesizing qualitative studies (Vol. 11): Sage Publications, Inc.

Obi, T. (2012). The 2012 Waseda University International e-Government Ranking. Retrieved 19 October, 2012, from http://www.cicc.or.jp/english/news/2012/e-GovRanking2012e.PDF

OECD. (2003). The e-Government imperative. Retrieved 3 October, 2012, from http://www.keepeek.com/Digital-Asset-Management/oecd/governance/the-e-government-imperative_9789264101197-en

Office of Management and Budget. (2005). Federal Enterprise Architecture Program : EA Assessment Framework 2. Retrieved 31 October, 2011, from http://www.cio.gov/Documents/OMB_EA_Assessment_Framework_2.pdf

Parasuraman, A., Berry, L. L., & Zeithaml, V. A. (1991). Refinement and Reassessment of the SERVQUAL Scale. *Journal of retailing*, 67(4), 420-420.

Parasuraman, A., Zeithaml, V., & Berry, L. (1985). A conceptual model of service quality and its implications for future research. *The Journal of Marketing*, 41-50.

Parasuraman, A., Zeithaml, V. A., & Berry, L. L. (1988). Servqual: A Multiple-Item Scale For Measuring Consumer Perceptions of Service Quality. *Journal of retailing*, 64(1), 12-12.

Pereira, C. M., & Sousa, P. (2005). Enterprise Architecture: Business and IT Alignment. *ACM Symposium on Applied Computing*, 1344-1345.

Peters, R. M., Janssen, M., & Engers, T. M. v. (2004). Measuring e-government impact: existing practices and shortcomings. Paper presented at the Proceedings of the 6th international conference on Electronic commerce, Delft, The Netherlands.

Petter, S., DeLone, W., & McLean, E. (2008). Measuring information systems success: models, dimensions, measures, and interrelationships. *European Journal of Information Systems*, 17(3), 236-263.

Pheng, T., & Boon, G. W. (2007). Enterprise Architecture in the Singapore Government. *Handbook of Enterprise Systems Architecture in Practice*, 129.

Pitt, L. F., Watson, R. T., & Kavan, C. B. (1995). Service quality: a measure of information systems effectiveness. *MIS Quarterly*, 173-187.

Ross, J. W., Weill, P., & Robertson, D. (2006). Enterprise architecture as strategy: Creating a foundation for business execution. Boston, MA: Harvard Business Press.

Saha, P. (2009). Architecting the connected government: practices and innovations in Singapore. *Paper presented at the IEGOV2009.*

Saha, P. (2010). Enterprise architecture as platform for connected government:

Advancing the Whole of Government Enterprise Architecture Adoption with Strategic (Systems) Thinking. Retrieved 3 October, 2011, from http://unpan1.un.org/intradoc/groups/public/documents/unpan/unpan041801.pdf

Seppanen, V., Heikkila, J., & Liimatainen, K. (2009, 20-23 July 2009). Key Issues in EA-Implementation: Case Study of Two Finnish Government Agencies. Paper presented at the Commerce and Enterprise Computing, 2009. CEC '09. *IEEE Conference on.*

Sessions, R. (2006). A better path to enterprise architectures. Microsoft Technical Articles, http://msdn2. microsoft. com/enus/library/aa479371. aspx,(accessed April 22, 2007).

Shah, H., & Kourdi, M. E. (2007). Frameworks for Enterprise Architecture. IT Professional, 9(5), 36-41. doi: 10.1109/mitp.2007.86.

Shahkooh, K. A., Saghafi, F., & Abdollahi, A. (2008, 7-11 April 2008). A Proposed Model for E-government Maturity. Paper presented at the Information and Communication Technologies: From Theory to Applications, 2008. ICTTA 2008. 3rd International Conference on.

Siau, K., & Long, Y. (2005). Synthesizing e-government stage models–a meta-synthesis based on meta-ethnography approach. *Industrial Management & Data Systems*, 105(4), 443-458.

Simeon, R. (1999). Evaluating domestic and international web-site strategies. *Internet Research*, 9(4), 297-308.

Stanforth, C. (2006). Using actor-network theory to analyze e-government implementation in developing countries. *Information technologies and international development*, 3(3), 35-60.

Tamm, T., Seddon, P. B., Shanks, G., & Reynolds, P. (2011). How Does Enterprise Architecture Add Value to Organisations? *Communications of the Association for Information Systems*, 28(1), 10.

The Open Group. (2009). TOGAF Version 9. Retrieved 9 September, 2011, from http://www.opengroup.org/togaf/

The United Nations. (2003). Global e-Government Survey 2003: E-government at the Crossroads. Retrieved 2 October, 2012, from http://unpan1.un.org/intradoc/groups/public/documents/ un/unpan016066.pdf

The United Nations. (2004). Global e-Government Readiness Report: Toward Access for Opportunity. Retrieved 2 October, 2012, from http://unpan1.un.org/intradoc/groups/public/documents/ un/unpan019207.pdf

The United Nations. (2005). Global e-Government Readiness Report 2005: From e-Government to e-Inclusion. Retrieved 2 October, 2012, from http://unpan1.un.org/intradoc/groups/public/ documents/un/unpan021888.pdf

The United Nations. (2008). UN e-Government Survey 2008: From e-Government to Connected Governance. Retrieved 2 October, 2012, from http://unpan1.un.org/intradoc/groups/public/ documents/un/unpan028607.pdf

The United Nations. (2010). United Nations E-Government Survey 2010: Leveraging e-government at a time of financial and economic crisis. Retrieved 2 October, 2012, from http://unpan1.un.org/ intradoc/groups/ public/documents/un/unpan038851.pdf

The United Nations. (2012). UN e-Government Survey 2012: e-Government for the People. Retrieved 2 October, 2012, from http://unpan1.un.org/ intradoc/groups/public/documents/un/unpan048065.pdf

Torres, L., Pina, V., & Royo, S. (2005). E-government and the transformation of public administrations in EU countries: Beyond NPM or just a second wave of reforms? *Online Information Review*, 29(5), 531-553.

Van Dyke, T. P., Kappelman, L. A., & Prybutok, V. R. (1997). Measuring information systems service quality: concerns on the use of the SERVQUAL questionnaire. *MIS Quarterly*, 195-208.

Venkatraman, N., Henderson, J. C., & Oldach, S. (1993). Continuous strategic alignment: exploiting information technology capabilities for competitive success. *European Management Journal*, 11(2), 139-149.

Watson, R. T., Pitt, L. F., & Kavan, C. B. (1998). Measuring information systems service quality: lessons from two longitudinal case studies. *MIS Quarterly,* 61-79.

Wegmann, A., Balabko, P., Lê, L. S., Regev, G., & Rychkova, I. (2005). A method and tool for business-IT alignment in enterprise architecture.

Wescott, C. G. (2001). E-Government in the Asia-pacific region. Asian Journal of Political Science, 9(2), 1-24.

Wu, R. C. Y. (2007). Enterprise integration in e-government. Transforming Government: People, *Process and Policy*, 1(1), 89-99.

Zachman, J. A. (1987). A framework for information systems architecture. IBM Systems Journal, 26(3), 454-470.

Zachman, J. A. (1997). Enterprise architecture: *The issue of the century. Database Programming and Design*, 10(3), 44-53.

Zachman, J. A. (2003). The Zachman Framework for Enterprise Architecture : A Primer for Enterprise Engineering and Manufacturing. Retrieved 5 July, 2011, from http://www.zachmaninternational.com

Zachman, J. A., & Sowa, J. F. (1992). Extending and Formalizing the Framework for Information Systems Architecture. *IBM Systems Journal*, 31(3), 590-616.

In: Technology Innovation Leadership … ISBN: 978-1-63117-388-2
Editors: G. Rampersad and F. Patel

Chapter 4

GREEN GEMS: LEADING CLEAN TECHNOLOGY INNOVATION NETWORKS

Karl Fradley[1] and Giselle Rampersad[2*]
[1]Environmental Protection Authority,
Government of South Australia, Australia
[2]Flinders University, Australia

ABSTRACT

In the race to innovate the green gems of the future, leading and investing in clean technology innovation offers a resounding call to action. For Middle Eastern oil-rich countries and energy companies seeking to diversify by investing in solar and wind power to others picking up the pieces after the inevitable closure of challenging sectors including automotive and coal, turning towards R&D and advanced manufacturing of these green gems offer a glimmer of hope. Such green gems range from solar powered cars and alternative energy solutions to sensor-led systems to recycle water, monitor air and water pollution and promote smart energy use in play and work. This chapter offers a case study of best practice, providing an example of how a clean technology innovation network can emerge comprising key actors from government and business to create clean technology solutions to address water pollution compliance legislation. New legislation coupled with the pivotal role of a green information system to enforce compliance has resulted in

* Corresponding author: giselle.rampersad@flinders.edu.au

the development of an entire industry and export sector around water treatment systems. The chapter offers important implications for policy makers and managers in leading the development of clean technology innovation sectors.

Keywords: clean technology, green innovation, Green IS, actor network theory, sustainability, regulatory tool, innovation

INTRODUCTION

Leading clean technology innovation is a recent interest for Middle Eastern countries in diversifying and investing the wealth derived from the oil industry in the development of future sustainable industries. There is an emerging focus on renewable energy in the Arab States of the Gulf that comprise the Gulf Cooperation Council (Bahrain, Kuwait, Oman, Qatar, Saudi Arabia and United Arab Emirates (UAE)). Renewable solutions may be applicable in remote areas (such as the Saudi Arabian mountains) and isolated areas (such as the Bahraini islands) with substantial potential for 'solar energy in GSM (Global System for Mobile Communications), road lighting, cooling and desalination' (Konstantinos, Doukas and Psarras, 2006, p. 3724). Abu Dhabi in UAE had focused on "transforming oil wealth into renewable energy leadership and has set the long-term goal of a transition from a 20th Century, carbon-based economy into a 21st Century sustainable economy…[as evidenced by its development of] Masdar City, a planned carbon-neutral town of Abu Dhabi" (Reiche, 2010a, p. 378). Furthermore, Dubai has implemented a green building code (Reiche, 2010b). Additionally, due to rising domestic consumption, the Iranian government has invested in developing the renewable energy sector with the establishment of the Renewable Energy Organization of Iran within the Ministry of Energy (Aslani, Naaranoja and Zakeri, 2012). There is also attention being placed on solar power for seawater desalination in the Middle East given the need to improve water management with limited potable freshwater, growing populations, increased urbanization and industrialization (Trieb and Steinhagen, 2008).

To develop the green sectors, innovation networks are important. These networks refer to groups of government agencies, universities and businesses working together to develop creative solutions (Rampersad, Quester and Troshani, 2009). Within these networks, governments can pave the way for the development of green industries by developing appropriate legislation. An example of this is the Environment Protection Act (S.A. Government, 2003), which was created by the South Australian government in 1993 and was designed to protect the environment in many forms. Another example is the Water Quality Act of the United States (U.S. Government, 1997). Enforcement of this legislation is also critical and governments have developed green information systems (IS) as innovative mechanisms to assist in sustainability initiatives (Melville, 2010). Green IS can be pivotal in fostering compliance with legislation, analysing sensor data, transmitting to controllers and providing information to consumers, governments and managers in monitoring their emissions and waste, and increasing their energy efficiency (Watson, Boudreau, & Chen, 2010). The business community is also vital in these clean technology networks and developing appropriate technologies to ensure compliance with regulatory requirements such as treatment systems and smart technologies.

This chapter is based on a case study which involves the development of new environmental legislation coupled with the development of a Green IS to enforce compliance. This has resulted in the development of an entire industry through building capacity within the industry involving the birth of component manufacturers and export firms around water treatment systems. This particular case sheds important insights because it employed an innovative methodology for enrolling the network actors. The analysis of how this legislation was implemented provides valuable information about how future sustainability sectors can be developed. This legislation stems from environmental degradation of the River Murray – Australia's largest river which is 2,375 kilometres in length. Decades of heavy water usage by irrigators in the River Murray combined with years of drought and unsustainable river practices have drastically reduced the water quality within the Murray system. To help combat the pollution of the River Murray, through wastewater discharge into the waters, the Environment Protection Authority (EPA) of South Australia introduced a Code of Practice designed to improve the water quality of this valuable area.

To enforce the regulation, a new Green IS system was implemented. This system was necessary given the unmanageably vast amount of data being recorded from a large number of actors within the network leading to data duplication, convolution and data loss. In order to better sustain and manage the large volumes of regulatory data, the EPA implemented a Green IS, called the "Vessel Management System" (VMS), which is a web based relational database that serves in not only storing data, but automatically monitoring compliance with regulation, and subsequently generating communication for network actors and identifying errors within the data.

This case study is important to a range of stakeholders. The Green IS system has played a critical role in the enforcement of legislation and as a result, it has fostered the development of an industry in water technologies as many innovative products and services have since been introduced to reduce water pollution in compliance with legislation. Consequently, firms operating in the water technologies sector would be interested in this case study given the impact of Green IS on strategic trends in the industry. Water technologies has become a major export for South Australia at AUD$400M with projected growth to AUD$1B by 2015 (EDB, 2009) and therefore, this case study is pertinent as it is based on an exemplary sustainable technology industry in Australia. Furthermore, government agencies with a focus on sustainability would view this case study as significant in understanding the critical role of Green IS in the successful enforcement of sustainability legislation and industry development. The chapter is also important to technology firms, contractors and software companies with a strategic focus on clean technologies in their offerings as this case study including a vast number of 2135 actors demonstrates the relevance and dynamics of the water sustainability segment as a potential market opportunity.

This chapter uncovers how Green IS plays a significant role in solidifying the relationships between regulatory bodies such as government and private industry to create a collaborative network in environment protection. Networks between government and industry are crucial for achieving innovative outcomes (Rampersad et al., 2010a; Rampersad et al., 2010b). The chapter contributes to the literature by applying Actor Network Theory (ANT) to both the legislative process and Green IS to inform understanding of how sustainability initiatives can be introduced and the role that Green IS can play in the successful regulation of initiatives.

Theoretical Background

Actor network theory (ANT) is an approach to social theory that was first developed by Callon et al. (1986). It is a particularly valuable tool because it is a network theory that studies not only people, but also objects and organizations (Singleton & Michael, 1993). It explores how relationships are formed between different networks as opposed to studying why they originate. The central tenant of actor network theory is the concept of hybrid entities, that is, entities that simultaneously contain both human and non-human entities (Murdoch, 1997). For example, a software development firm contains many of the typical entities. It has managers, software designers, developers and many other social forces. Central to the firm, however, are non-human entities such as technical standards, compilers, software development tools and various computer programs. The absence of any of these non-human entities would prove disastrous, giving their significant roles within the overall network.

Actors are more difficult to define. An actor can be a simplification of a network. For example, one may regard a governmental regulatory body as an actor, whilst analysed much more closely it is a network of many other interrelated actors. To quote Callon (1987), "[an actor] ... can be compared to a black-box that contains a network of black-boxes that depend on one another both for their proper functioning and for the proper functioning of the network" (Callon, 1987, p. 95).

Translation is another central concept within ANT. Simply defined, translation is "... the means by which one entity gives a role to others" (Singleton & Michael, 1993, p. 229). The process by which translation happens can be divided up into four key phases (Callon, 1986). The first phase is problematization. This phase involves defining what problems need to be solved and defining the relevant actors involved. Delegates, that will represent groups of actors, need to be identified. In this phase, there is a primary actor who will try and establish itself as an obligatory central passage point between the other actors and the network. Interessement, the second phase, focuses on generating interest among the other actors in the areas of interest and goals of the primary actor.

It is in this phase that the primary actor will attempt to convince the other actors that the roles they have been given are acceptable and their terms of involvement will be negotiated. Enrolment, the third phase, is where the actors that have had their roles defined in the interessement stage accept their position within the network. The successful formation of networks does not bring the process to a close.

Networks within ANT are often described as transient, in that they have the potential to be constantly changing. New actors may enter the network, existing actors may leave, alliances between entities can fracture and networks themselves can become unstable and collapse, only to reform again in a different configuration (Callon, 1986). The fourth and final phase is the mobilisation of allies. This is the phase where, assuming the delegate actors in the networks sufficiently represent the masses, enrolment becomes active support.

While ANT has been applied in various contexts, there is a scarcity of research on ANT in a Green IS context. Despite its critics (Doolin & Lowe, 2002; Walsham & Sahay, 1999), ANT has been employed in various contexts including exploration of the England-based Luddite movement (Grint & Woolgar, 1997); process analysis of scientists (Latour, 1987); the invention of the Kodak camera (Latour, 1991); the workings of the France-based transportation system (Latour, 1996); understanding users as social actors (Lamb & Kling, 2003); and evaluating interpretative field studies in IS (Kline & Myers, 1997).

More recently it has begun to be applied in the Green Information Technology (IT) context in a study undertaken by Bengtsson and Agerfalk (2010).

This chapter is a direct response to Bengtsson and Agerfalk's (2010) calls for research on ANT in various sustainability contexts beyond the context of Swedish transportation IT undertaken in that study. This chapter is based on an in-depth case study on the use of Green IS in a water regulation compliance context in Australia. Unlike any previous study it contributes by examining the use of ANT in understanding the successful community integrated approach and the effective use of Green IS in the regulation and enforcement of sustainability initiatives.

Actor network theory has been chosen as the theoretical frame for this chapter because it provides an effective tool to analyse an innovation such as Green IS that affects many entities.

There are two types of innovation, incremental and emergent. Incremental innovation is where a product is improved in a step-by-step process, whilst an emergent innovation is where an entirely new product or feature is introduced. McMaster et al. (1997) contends that innovations are created "... from chains of weaker to stronger associations of human and non-human alliances. ... Each actant translates and contributes its own resources to the shape and ultimate form of the emerging black box" (McMaster et al., 1997, p. 4). In order for an innovation to be successful, it needs a consortium of actors powerful enough to drive this process.

A failure with regards to the adoption of the innovation can be viewed as a failure of those involved for not being able to construct a strong enough network of alliances between the relevant actors (Tatanall and Gilding 1999, McMaster et al 1997). In order to make an innovation successful, the enrolment of others is important.

Case Study

Case studies have been an important facet of information systems research as it is appropriate for the study of information systems within and across network actors (Kline & Myers, 1997; Lee, 1989; Orlikowski & Baroudi, 1991). This case study is based on a Green IS monitoring system for Australia's largest river, the River Murray, given the significant interest to both government and industry actors. Currently, there are thousands of vessels that operate both commercially and privately in the South Australian section of the River Murray. These vessels can range from multi-storey luxury commercial houseboats to modest vessels used for holiday homes. The vast majority of these vessels contain basic facilities such as toilets and showers, whilst the larger vessels contain facilities such as spa baths, washing machines and multiple galleys. Toilets produce blackwater, which is comprised largely of sewage whilst galleys, dish washers and other facilities produce greywater which is largely comprised of food waste, soaps and other cleaning products. The standard practice for the past few decades has been to simply discharge all of this waste into the River Murray. This created many environmental issues. Discharges were often observed by many members of the public in areas such as marinas and swimming areas and the wastewater would contaminate waters where humans would swim. In addition, the discharges presented numerous water quality issues such as increased bacterial counts, suspended solids, E.Coli, grease and oil (DeIonno & Thomas 2009). In addition to vessels, the facilities that vessels would use for maintenance, refuelling and other facilities have been identified as harmful to the environment.

The South Australian Environment Protection Authority (EPA), observing the harm being done to the environment through wastewater discharged, took action to protect South Australian waters. The central actor for this case study is the South Australian EPA, who is responsible for the introduction and regulation of the new legislation designed to prevent wastewater discharges. Further network actors are other government departments, such as the Department for Transport and Infrastructure and the Department of Environment and Natural Resources.

Another key network actor is industry, or every non-government entity that would be affected by the introduction of the legislation. This actor is comprised of many other actors including numerous commercial vessels, privately owned vessels, boating associations, yacht associations and various others such as marinas.

The Vessel Management System

The Green IS, the Vessel Management System, a web-based system with a centralised database, was introduced to coordinate the information and formalise relationships between the numerous network actors. The system facilitated the instantaneous access of all vessels in a given month that were involved in a compliant, and it subsequently facilitated contact with the actors to ensure compliance with legislation. Overall the system lead to more efficient use of the EPA's time and resources and it became central to the regulation of environment protection legislation.

The EPA identified its information management as an important area for improvement. As the EPA enforced the Code of Practice, it collected a vast array of data about the vessels falling under the umbrella of the legislation. Information about vessels included the facilities, technical specifications of vessels and other pieces of information. In addition to the vessels was information about the owners including any correspondence and action between the owners and various government and industry players that was of relevance to the regulation. The EPA was also trying to track down previous owners of vessels and maintain information about the introduction and performance of greywater treatment systems.

With the regulation of thousands of vessels underway, the amount of data collected became increasingly large. The method being used at the time was multiple spreadsheets for different type of vessels, for example commercial and marine vessels had their own spreadsheets whilst private vessels and audits were stored in another. The sheer volume of data stored in these spreadsheets caused the spreadsheets to become unmanageable. In addition some vessels would move between commercial and private statuses causing double ups in data along with an assortment of other errors.

This created difficulties for the regulators with regards to enforcing the environment protection legislation on the industry and represented a weakness in the network. After several discussion meetings and research into the regulator needs of the organization it was decided that a new, relational information system was necessary.

The solution was what became known as the Vessel Management System. The interconnected set of skills and competencies comprising the innovation conceptualised and then developed the database as a solution to all of the issues.

The introduction of this system had a significant impact on the network. The vessel management system allowed for the proper management of correspondence, along with documents related to vessels and photos. This allowed for the EPA to be more aware of not only a vessel's current situation, both with regards to compliance and to technical specification but also to be aware of the history of the vessel. With a structured correspondence management system in place, each environment protection officer was aware of the correspondence with other officers.

Efficiencies were gained in the following areas:

- A means to track the previous owners of vessels and compliance history across time
- Automatic identification of vessels with upcoming compliance dates
- For the first time, statistics of all compliance figures were made available allowing Environment Protection Officers to become more aware of the impact their work is having
- Elimination of duplicate records and cleaner data
- A centralised repository comprised of all data spreadsheets, physical checklist and hardcopy reports obtained from other departments
- Automatic generation of letters as opposed to lengthy mail merge processes
- Photo galleries for each vessel in the database

Whilst the innovation system had no direct interaction with the industry actor, the effects of it could be felt through the regulatory body. With a greater awareness of forthcoming compliance dates and the ability to automatically generate letters to be sent out to vessel operators, the process of regulation became far more streamlined. With more efficient regulation underway, this lead to a greater amount of vessels becoming compliant and subsequently meant that both the Code of Practice was being upheld and that the inland and marine waters were being better protected.

RESULTS

To examine the development of the sector and the pivotal role of the Green IS, this chapter identifies network actors as the process of translation unfolds. The result was the network in Figure 1 which emerged as translation occurred. A total of 2135 actors were identified within the network (EPA, 2008).

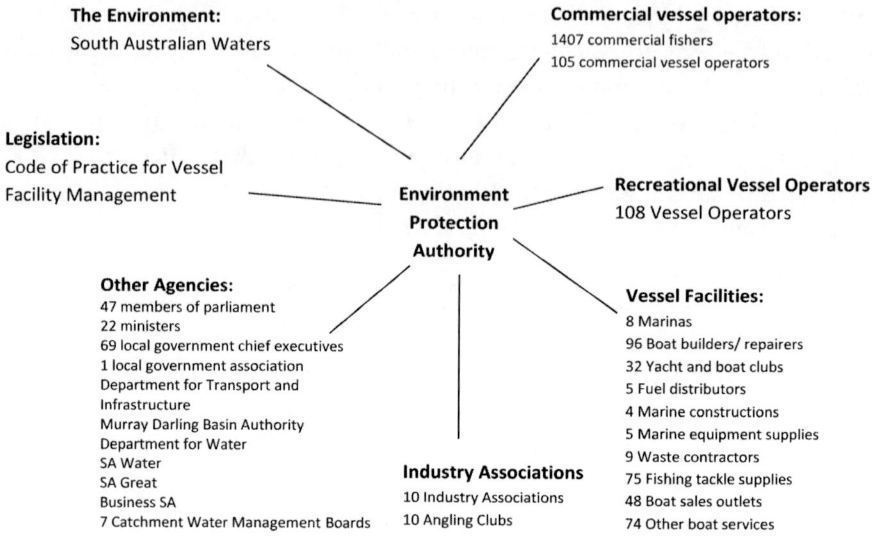

Figure 1. Green IS sustainability monitoring network.

Problematization

Complaints about wastewater discharge, particularly in vicinities used by the public for recreational purposes, were received by the State government, who was already aware of vessel discharge. Before any response could be made the government had to identify what damage was occurring. The first question was simple, 'what was the harm of the wastewater discharge into the inland and marine waters'? In addition to this question came other inquiries. What other activities taking place in inland and marine waters were causing environmental harm? How could the environmental harm be minimized or eliminated?

A risk assessment was conducted. This assessment consisted of the analysis of studies conducted from around the world. Furthermore, the government utilized its resources to conduct studies and consulted with members of the public. There was consensus that discharges were harmful to both the environment and potentially to recreational users of the river. This gave rise to numerous other questions. What other actors would be potentially affected by this network? Who are the additional actors? How do you enroll actors into a policy that would cost them significant amounts of money?

Should all wastewater discharge be prevented or could there be another method? Many more questions were postulated during this phase.

Throughout the problematization phase the members of the EPA developed one specific argument: that the inland and marine waters of South Australia should be protected. As Figure 1 demonstrates, there were a large number of actors that would be affected by the conceptualized Code of Practice. In addition to the environmental harm identified from wastewater discharge, continued investigation into vessels and vessel facilities in waters identified a number of other activities of potential environmental harm. These included (Ballantine, 2008):

- Vessel facility operation
- Dredging
- Maintenance areas (such as slipways and hardstands)
- Vessel cleaning
- Abrasive blasting, scraping and sanding
- Pressure water blasting
- Fibreglassing
- Painting and varnishing
- Welding and metal fabrication
- Engine maintenance and repair works
- Refueling
- Wastewater for marine and inland waters
- Bilge water
- Biofouling
- Ballast water
- Vessel recovery

The EPA, in conjunction with both the community and other departments, created a state wide stakeholder database. This database stored a comprehensive list of stakeholders that was derived from the work of the External Advisory Group combined with an extraction of business listings from the yellow pages. The following list is adapted from the stakeholder database.

South Australian Waters: the inland and marine waters that the EPA is wishing to protect from the harm of wastewater discharges. These waters are central to the commercial operators, in that their businesses rely on the water. Fishermen require healthy waters so that fish may survive. Houseboat operators need clean waters so that people may enjoy recreational activities within. Vessel facilities require these waters to be in good condition, so that their businesses can operate and tend to the needs of the vessels.

Code of Practice for Vessel Facility Management (Inland and Marine Waters): the goal of the central actor, the EPA is to protect the health of the River Murray. This piece of legislation was their solution for promoting sustainability. The concept of the code would impact on all of the other actors in order to protect the South Australian waters.

Other Government Sectors: identified in this category were 47 members of parliament, 22 ministers and 69 local government chief executives. Government departments such as the Department for Transport and Infrastructure, Murray Darling Basin Authority and the Department for Water were identified, amongst many others, as key stakeholders.

Commercial Vessel Operators: use the waters to operate their businesses. They can range from fishermen to companies offering a vacation aboard a luxury a houseboat. They would be required to make adjustments to their vessel, in accordance with any new legislation, that would prove to be very costly, making the idea unattractive.

Recreational Vessel Operators: which include people that live aboard houseboats and possess such vessels for holiday purposes.

Vessel Facilities (Marinas/Slipways/Yacht and Boat Clubs): 48 boat sales outlets were identified. They would be affected by the regulations as boats constructed in the future would have to be compliant with any introduced legislation. There were 8 marina operators, 96 boat builder repairers, 32 yacht clubs and numerous other groups such as various boat services, waste contractors, yachting and boating clubs.

The process of problematization identified not only the actors but their relationships to each other within the network of industry. Vessels may discharge wastewaters into water bodies, but the facilities used by vessels also caused environmental harm. An example of this can be found with vessels that have engine repairs. Marinas and slipways, whilst making engine repairs, would accidentally discharge small amounts of oily substances into the water bodies resulting in environmental harm.

Interessement

The interessement phase occurs when a central entity influences and secures the identification of actors that emerged during the problematization phase (Callon, 1986). Literally, it is about getting the other actors within the network interested in conforming to the goals of the central actors. The goal of the EPA was the prevention or minimization of all environmental harm resulting from vessels and vessel facilities.

For example, the agenda of a typical recreational boat owner can range from a relaxing vacation to enjoying a life living aboard a vessel. In the past, few have been concerned about the wastewater discharges and the polluting action had become a part of standard practice. During interessement, the EPA would ensure that these boat operators – along with the other previously identified actors – are interested in taking personal actions with regards to reducing the environmental impacts of their own personal vessels.

This phase revolves around convincing actors that the waters are worth protecting and need to be protected. It focuses on demonstrating to the relevant actors that their own practices are causing harm to natural resources that have become a central part of their business and daily life.

To do this, the EPA used an innovative approach. An external advisory group, comprised of key stakeholders including business, government and association figures was created. This group was formed to give input on the development of the Code of Practice and to try and create a set of guidelines that would both be useful for the businesses and also environmentally sustainable. The EPA utilized this group to provide information to the relevant stakeholders, showing them the impact that wastewater was having on the environment.

In addition to the advisory group, information on the Code of Practice was released to the media and presented to the public in the form of printed advertisements, media releases and media contracts. Members of the public and anyone affected by the code were invited to consultation forums to give their opinions about the code.

The EPA utilized the media and advisory groups to put forward the stance that everyone who discharged wastewater needed to either stop or reduce the harm resulting from the discharges. It assigned role of environment protection to everyone.

Furthermore, the EPA sent out letters announcing the formal consultation process of the draft version of the Code of Practice. These letters provided interested actors information on where copies of the draft could be obtained, gave invitations to the public meetings and workshops, provided supporting documentation outline about how interested parties could comment on the draft.

Enrolment

Enrolment is where the other actors in the network accept their role. Private vessel owners accept that their practices are harming the environment and take corrective actions. This phase was about negotiation and determining the appropriate balance between environmental goals and community goals.

The consultation and submissions programs allowed the industry and affected parties to have a voice. Over the course of numerous consultation meetings areas of change within the initial drafts of the legislation were identified. The EPA made technical modifications to the code to better suit the community whilst still addressing the core issue of environment protection.

Furthermore, submissions from various people and organizations resulted in numerous changes and re-drafting of the code to create something that worked for the vast majority of stakeholders. As new drafts were produced the consultation process continued, more aspects were identified for environment protection whilst other restrictions were eased off to better suit the needs of stakeholders.

The EPA essentially attained enrolment of the other actors through co-development of the legislation. Through integrating the community in the development, few could argue that the government was insensitive of industry and private entities. Through small changes to the codes the actors that would be responsible for preventing wastewater pollution were in agreement with the legislation and were more willing to adopt sustainable practices in their activities.

Mobilisation of Allies

The previous phases outlined who was identified as representatives for the various actors within the network. These representatives, in previous phases, were responsible for the agendas of those entities that they represented. For instance, representatives of the industry organization were speaking on behalf of their respective organizations. This begged the question, would the represented follow the representatives? Are the spokesmen representatives of the masses? Whilst the representatives spoke on behalf of those they represented, the represented themselves were encouraged to make submissions and discuss their own personal views. Through allowing the represented to have their own voices, the EPA could properly determine whether or not the spokespeople were indeed representative.

One of the major advantages of actor network theory is the ability to analyze the effects that each actor has on one another. The industry network, as previously mentioned is in and of itself a network consisting of many individual entities. Vessels utilize vessel facilities such as slipways to conduct modifications on their vessels and vessel owners can be members of clubs and associations. With the enrolment of some of the major associations, came the enrolment of the members of the associations. As the sustainability initiative gained more and more enrolment, the mobilization of allies resulted in the industry furthering the cause of the EPA simply through their own enrolment. As the government offered rebates to vessel operators such as boat owners to participate in trials of grey water treatment systems, it lead to a demand for and development of clean technologies around grease removal, nutrient removal, fine and dissolved, solids removal, sterilization, de-chlorination, waste management and disposal.

As with any significant piece of legislation, there was opposition. Because of the community integrated approach to the development of the Code of Practice, the opposition was localized and mostly limited to certain individual operators and some commercial enterprises.

Network Reconfiguration

With the legislation introduced, the EPA's role changed from the introduction of the sustainability initiative to the regulators of it. The EPA would subsequently be responsible for ensuring that vessels and facilities were in fact complying with the legislation. It is in this regulation phase that Green IS had a crucial role.

To properly analyze the role of Green IS in the regulation of the sustainability initiative, this chapter will utilize a modified framework that is based on actor network theory (Lyytinen & King, 2002). This framework, first developed by Lytinen and King (2002) then utilized and applied to mobile services (Yoo, Lytinen, & Yang, 2005) describes three main realms. These three realms are the innovation systems, the marketplace, the regulator regime and standards, policies, directives, incentives as illustrated in Figure 2.

To better understand the impacts of the Green IS within the network, Figure 2 examines the relationships among the various entities in the network that were previously identified in Figure 1. In this instance, the standards, policies and directives are embodied by the Code of Practice for Vessel Facility Management. This Code of Practice includes policies that are set out to protect inland and marine waters, as such, the inland and marine waters are a key actor within this configuration. The waters are used by the industry, but to continue using these waters they must follow the legislative rules outlined in the Code of Practice. As such the relationship between the Code of Practice and the environment is so tightly linked that they are connected and become a hybrid entity. Regulators regulate industry by enforcing the code of practice to protect inland the marine waters. The Green IS system is used to inform regulators and monitor industry compliance with legislation enforces by regulators.

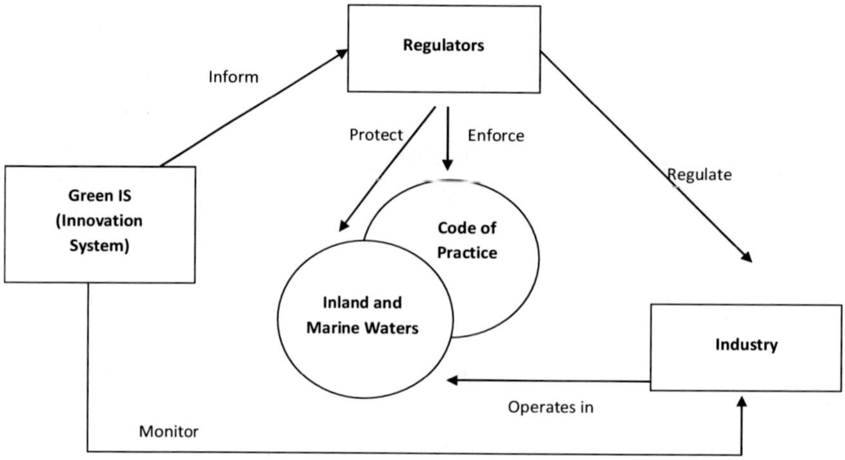

Figure 2. Network reconfiguration and the role of Green IS.

Conclusion

The chapter contributes by examining the development of an innovative clean technology sector by coupling legislation with a Green IS for ensuring compliance. It highlights the critical role of the Green IS by examining the successful implementation of innovative legislation in a water sustainability context.

This chapter may be of interest to a range of stakeholders. Governments charged with the responsibility for sustainability compliance may find this case study as a useful example of how a clean technology sector can be developed. Firms operating in the clean technology sector for instance water technologies would find this case study important in recognising opportunities for market development surrounding compliance with legislation. Software companies and IT managers focused on the development of Green IS would find this chapter insightful in identifying a successful application for Green IS in water sustainability initiatives.

While this chapter offers important managerial implications, it should be interpreted in light of its limitations. It was conducted on one case study in a network surrounding water pollution in an Australian context and therefore further research is necessary in other legislative contexts. Future studies can also be based on longitudinal research in these networks (Plewa et al., 2013) that examines the reduction of pollution from these initiatives. Nevertheless, the network under investigation in this chapter is very extensive comprising of 2135 actors. The chapter is significant as it provides a rich and in-depth case study demonstrating how governments can drive the development of clean technology innovation networks through legislation and enforcement through technology.

REFERENCES

Aslani, A., Naaronja, M. & Zakeri, B. (2012). The prime criteria for private sector participation in renewable energy investment in the Middle East (case study: Iran). *Renewable and Sustainable Energy Reviews.* 16(4), 1977-1987.

Bengtsson, F., & Agerfalk, P. (2010). Information technology as a change actant in sustainability innovation: Insights from Uppsala. *Journal of Strategic Information Systems, In-Press (DOI:10.1016/j.jsis.2010.09.007).*

Blowfield, M., & Visser, W. (2007). *Sustainability Innovation: Mapping the Territory.* Cambridge, United Kingdom: Cambridge University Press.

Callon, M. (1986). *Some elements of a sociology of translation: domestication of the scallops and the fishermen of St Brieuc Bay.*

Callon, M. (1987). Society in the making: The study of technology as a tool for sociological analysis. In W. E. Bijker, T. P. Hughes & T. P. Pinch (Eds.), *The social construction of technological systems* (pp. 85-103). Cambridge MA: The MIT Press.

Doolin, B., & Lowe, A. (2002). To reveal is to critique: actor-network theory and critical information systems research. *Journal of Information Technology, 17,* 69-78.

EDB. (2009). Economic Statement: South Australia's Prospects for Growth. A Report by the Economic Development Board.

EPA. (2008). South Australian EPA Code of Practice for Vessel and Facility Management: Marine and Inland waters: An Overview of Consultation: Environment Protection Authority

Freeman, C. (1987). *Technology policy and economic performance: lessons from Japan*. London, United Kingdom: Pinter.

Government, S. A. (2003). Environment Protection Act 1993.

Government, U. S. (1997). Water Quality Act 1997.

Grint, K., & Woolgar, S. (1997). *The Machine at Work – Technology, Work and Organization*. Cambridge: Polity Press.

Kline, K., & Myers, M. D. (1997). Qualitative research in information systems. *MIS Quarterly, 21*(2), 241-258.

Konstantinos, D.P., Doukas, H., and Psarras, J. (2006) Enhancing renewable energy in the Arab States of the Gulf: Constraints & efforts. *Energy Policy*. 34, 3719-3726.

Lamb, R., & Kling, R. (2003). Reconceptualizing users as social actors in information systems research *MIS Quarterly, 27*(2).

Latour, B. (1987). *Science in Action: How to Follow Engineers and Scientists Through Society*. Milton Keynes: Open University Press.

Latour, B. (1991). Technology is society made durable. . In J. Law (Ed.), *A Sociology of Monsters, Essays on Power, Technology and Domination* (pp. 103-131). London: Routledge.

Latour, B. (1996). *Aramis or the Love of Technology*. Cambridge, Ma.: Harvard University Press.

Lee, A. S. (1989). A scientific methodology for MIS case studies. *MIS Quarterly, 13*(1), 33-50.

Lyytinen, K., & King, J. L. (2002). Around the cradle of the wireless revolution: the emergence and evolution of cellular telephony. *Telecommunications Policy 27*, 97-100.

Melville, N. P. (2010). Information systems innovation for environmental sustainability. *MIS Quarterly, 34*(1), 1-21.

Murdoch, J. (1997). Inhuman/nonhuman/human: actor-network theory and the prospects for a nondualistic and symmetrical perspective on nature and society. *Environment and Planning: Society and Space 15*(6), 731 – 756.

Orlikowski, W. J., & Baroudi, J. J. (1991). Studying information technology in organizations: Research approaches and assumptions. *Information Systems Research, 21*(1), 1-28.

Plewa, C., Baaken, T., Macpherson, G., Korff, N., Johnson, C. and Rampersad, G.C. (2013). The Evolution of University-Industry Linkages:

A Framework. *Journal of Engineering and Technology Management*, 30 (1), 21-44.

Rampersad, G., Quester, P. & Troshani, I. (2009). Management of networks involving technology transfer from public to private sector: a conceptual framework. I*nternational Journal of Technology Transfer and Commercialisation,* 8 (2/3), 121-41.

Rampersad, G. C., Quester, P. & Troshani, I. (2010a). Examining network factors: commitment, trust, coordination and harmony. *Journal of Business and Industrial Marketing,* 25 (7), 487-500.

Rampersad, G. C., Quester, P. & Troshani, I. (2010b). Managing innovation networks: Exploratory evidence from ICT, biotechnology and nanotechnology networks. *Industrial Marketing Management*, 39 (5), 793-805.

Reiche, D. (2010a) Renewable energy policies in the Gulf countries: A case study of the carbon-neutral "Masdar City" in Abu Dhabi. Energy Policy. 38(1), 378-382.

Reiche, D. (2010b) Energy policies of Gulf Cooperation Council (GCC) countries-possibilities and limitations of ecological modernization in rentier states. *Energy Policy.* 28(5), 2395-2403.

Singleton, V., & Michael, M. (1993). Actor-Networks and ambielence: General practitioners in the UK cervical screening programme. *Social Studies of Science, 23*, 227-264.

Walsham, G., & Sahay, S. (1999). GIS for district-level administration in India: problems and opportunities. *MIS Quarterly, 23*(1), 39-66.

Watson, R. T., Boudreau, M. C., & Chen, A. J. (2010). Information systems and environmentally sustainable development: Energy informatics and new directions for the is community. *MIS Quarterly, 34*(1), 23-38.

Yoo, Y., Lytinen, K., & Yang, H. (2005). The Role of standards in innovation and diffusion of boradband mobile services: the case of South Korea. *Journal of Strategic Information Systems 14*, 323-353.

In: Technology Innovation Leadership ... ISBN: 978-1-63117-388-2
Editors: G. Rampersad and F. Patel

Chapter 5

THE IMPACT OF USING SOCIAL NETWORKING SITES (SNS) IN SMALL AND MEDIUM ENTERPRISES (SMEs) COMPARED TO LARGE ENTERPRISES

Rowa Barashi*
Flinders University, Australia

ABSTRACT

With the rapid change of technology, the use of Social Networking Sites (SNS) has been extended beyond individuals to professionals and organizations. These latter groups use SNS to engage with their target audiences. Yet, research about the impact of SNS on small and medium enterprises (SMEs) compared to large enterprises is limited. The purpose of this chapter is to examine the differences between SMEs and large enterprises regarding their use of SNS. Based on data from both Saudi Arabia and Australia, it will investigate the usage, objectives, management, benefit, and barriers of using SNS for firms of varying sizes.

* rbrashi@hotmail.com

INTRODUCTION

Over the past few years, the usage of Social Networking Sites (SNS) has increased. A diverse set of social media tools has been used, such as Facebook, Twitter, LinkedIn, YouTube, Slide Share, Flickr, and Google+. The essential service that the SNS provide is the extensive connectivity between individuals which lead to organizations considering them as a valuable marketing tool (Geho & Dangelo, 2012). SNS have facilitated increased customer engagement with businesses. This engagement impacts on the reputation and the brand of the company and in turn on its profitability (Quashi, 2009; Waters et al., 2009).

More recently, organizations have recognized the usefulness of social media as intrinsic in their marketing efforts as the business cannot be sustained with traditional marketing tools. The use of social media can contribute towards building strong relationships with customers. It is particularly suitable for small and medium sized enterprises (SMEs) because of its benefits, such as low cost and low level of IT skills required. Although large enterprises have also tried to use social media as a tool to support their business to foster loyalty and brand awareness and to follow the rapid change of technology, they have done so less successfully that their smaller counterparts (Derham et al., 2011).

Many researchers have established the benefit and the impact of social media on SMEs and how organizations use SNS to achieve brand objectives. SMEs and large enterprises have different objectives for using the SNS and face different issues. Yet research remains limited on what is the impact of social media on SMEs and how they are different in implementation compared to large enterprises in terms of their objectives, usage, barriers, challenges and opportunities.

The aim of this chapter is to address the gap by analyzing the current usage of SNS in SMEs as well as large enterprises. More specifically, the main research question of this study is: What is the impact of social media on SMEs compared to large enterprises?

LITERATURE REVIEW

A decade ago, the use of the Internet predominantly focused on reading, watching, and buying goods. More recently, the use of the Internet is increasingly sophisticated through web-based technology that create interactive platforms such as blogs, social sites and Wikipedia (Eileen & Rebecca, 2011; Waters et al., 2009). These platforms were created to share and discuss Internet content (Kietzmann et al., 2011). These tools and services were established many years ago after the Internet was launched but SNS were unknown to many users at that time.

Social Networking Sites

A definition of Social Network Sites is provided by Nielsen-Wire (2010): "SNS can be defined as virtual collections of user profiles which can be shared with others". Currently there are more than 150 SNS. The largest SNS is Facebook, followed by Twitter, LinkedIn, MySpace, and Google+ (eBizMBA, 2012). These social sites have different building blocks, scopes and functionalities.

Social media and Organization

SNS have a powerful impact on businesses, although there is still some confusion among managers, executives and academic researchers. Some executives avoid this form of media and ignore the opportunities and the challenges represented by social sites (Berthon et al., 2007). Many reasons behind this behaviour include the lack of understanding of the meaning of social media (Kaplan & Haenlein, 2010). The second reason is that some businesses think that social media is just for personal use. In addition, they fear that they may not be able to control customers' comments and they have a lack of expectation of what might be written. Besides this, some companies find it difficult for staff to follow the rapid change of technology, especially by "old employees" who are not familiar with recent technology, and who would have difficulty using the appropriate SNS or combination of them (DiStaso et al., 2011).

Various companies currently use SNS in order to strengthen their brand (Muniz Muniz & O'Guinn, 2001) and some for marketing research (Kozinets, 2002). The majority of these businesses have focused on SNS as a marketing tool and little is known about the impact of social sites on business (Eileen & Rebecca, 2011)

Objectives and Usage of SNS

Introducing SNS into an organization is influenced by objectives. In business, it is critical to specify the objectives of the company and how SNS can assist in achieving them (Hoffman & Fodor, 2010). Using a set of SNS tools can be beneficial to the company. However, without clear objectives, the company will not implement SNS effectively (Mangiuc, 2009).

Policies and Principles

Social media policies and guidelines are critical considerations when using social media as a tool in the organization, regardless the size of the company. Organizations should create their own policies and principles to monitor and build the structure of communication between the customers and the company relating to the use of SNS and can pertain to their employees, legal issues, security, privacy, regulation, monitoring and content management (Hurd, 2012).

Benefits / Obstacles

The benefits that some organizations face might be challenges or difficulties for others; all are based on the objectives achieved and the size of the industry. In general, SNS have the potential to benefit firms of all sizes. Firms use SNS, among other things, in order to increase traffic to its website, build relationships between the company and its users, build a new community, identify market opportunities, share information, interact with the feedback from customers and build brand power (Breslauer & Smith, 2009; e-Marketer, 2010). Despite benefits of SNS, obstacles also arise such as a lack of money, time, training, knowledge and being unfamiliar with the new technology (Buehrer et al., 2005; Venkatesh & Davis, 2000).

Methodology

Due to little prior research examining the impact of SNS on SMEs compared to large organizations, this research is exploratory. It is based on a mixed method approach combining both qualitative and quantitative methods.

First, a qualitative research method has been used. The reason behind this is the need for deep understanding of the situation (Yin, 2009). Data was collected via face-to-face interviews with 20 different enterprises, all of whom have adopted social media. They vary widely in terms of their levels of usage and the types of businesses and size. Second, a questionnaire was used to triangulate findings. The questionnaire was distributed to companies electronically through email and SNS such as Facebook and Twitter. The questionnaire was divided into categories:

- General Questions
- Social Networking sites
- Adoption Time
- Management
- Measurement
- Benefits – Opportunities
- Barriers – Challenges

The questionnaire was dsitributed based on a random sample. The sample was 82 companies, of which 15 companies responded effectively. The total responses from Australian companies were 7 companies and 8 were from Saudi Arabia. These companies were from different industry sizes. 6 companies were small, 5 were medium and 4 were large enterprises. Results were received within a two-month period in Saudi Arabia and Australia in 2012.

RESULTS

Figure 1 illustrates the uses of SNS by respondents highlighting the use of Facebook then twitter was the most popular SNS among all types of companies surveyed.

Table 1 summarizes the objectives of using SNS among small, medium and large enterprises.

Adoption Time

Figure 2 shows the length of time spent by businesses in adopting SNS. It also highlights the differences between organizations.

Table 1. The objectives of using SNS

	Small enterprises	**Medium enterprises**	**Large enterprises**
Objectives	To gain more customers To follow what the customers use To reach target audiences To gain popularity To interact with suppliers To promote brand	To follow what the users use To gain more customers To gain profit To build awareness To create presence in major channels To enhance customer service plans.	To engage with the community To be more accessible to customers To build relationships with consumers To listen to customers To contribute in the overall marketing strategy To achieve customer excellence To create channels with top media influencers To build awareness and loyalty To enhance search engine optimisation (SEO)

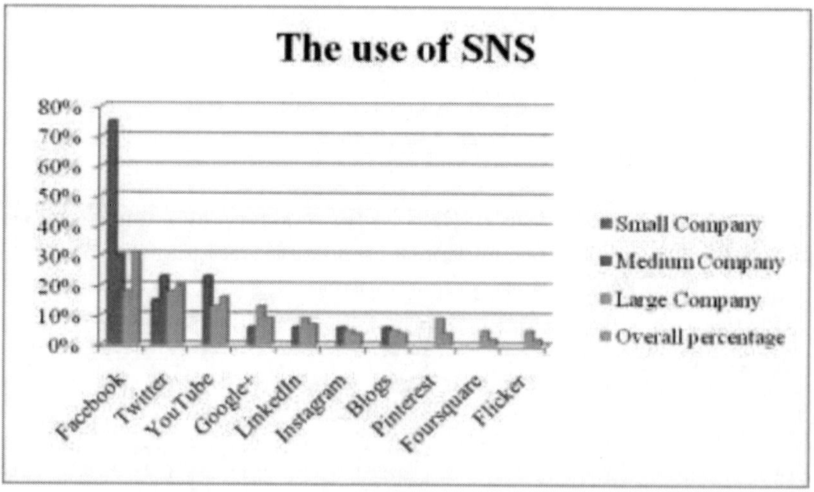

Figure 1. Social Networking Site: The Use of SNS.

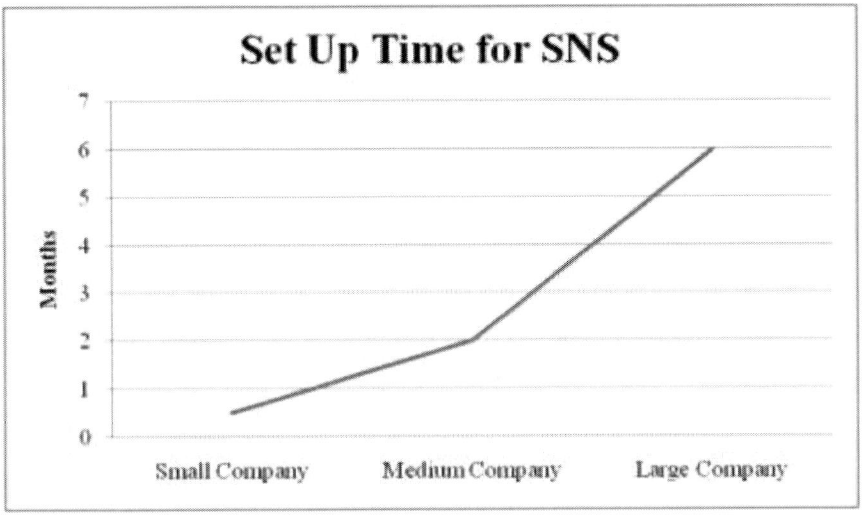

Figure 2. Adoption Time: SNS set up time.

The time for setting up grew steadily from 2 weeks to 1 month in small enterprises, from 2 to 5 months in medium enterprises and as the size increases the time for setting up rose to more than 6 months for large companies.

Policy, Strategy and Guidelines

Most of the small enterprises interviewed said that they did not have any specific policies around social networking sites. Medium enterprises on the other hand had some policies including the promotion of products and services and the dissemination of news, only following formal authorization. Finally, large enterprises had set SNS policies both directed towards employees and customers.

Management Role

Management played a significant role in SNS adoption. In most cases, the degree of positive participation of the company's employees depended on Management's encouragement. It was essential that Management understood the implications of social media and invested in its success.

Dealing with Comments, Suggestions and Complaints

Managing comments and suggestions is one role of a social media team and it can be done by trying to answer as many questions as possible on a timely basis. Most of the small companies indicated that they attempted to keep the page for positive activities rather than open suggestions. Medium sized enterprises dealt with it differently. They addressed comments and enquiries by consumers personally, not in the SNS. Large enterprises used a policy to deal with unsatisfied customers, sometimes the marketing team responded personally to complaints.

Measurement

This section illustrates how companies measure the effectiveness of using SNS. This is significant in answering one of the research questions. Small companies assessed SNS use by the increase in the number of customers, suggestions, and comments such as “likes”, “check ins”, “talking about”.

Return on Investment (ROI) is hard to measure for social media in medium enterprises.

They used the “Hootsuite” tool for customer services and brand monitoring and various tools for twitter analysis. Large enterprises used measurement tools such as: analytics, including Facebook Insights, Google Analytics looking, and TweetReach.

Benefits – Opportunities

The advantages of using SNS in a small, medium and large company were illustrated in Table 1.

The opportunities included advertising to other suppliers, and building customer networks by friends of friends and Electronic Word Of Mouth (E-WOM).

Barriers - Challenges

The last part of the questionnaire was about the barriers facing companies due to SNS. The fear and threats are different in small, medium, and large businesses.

Most of the small companies complained that time was one of the biggest barriers of using SNS followed by the misuse and mismanagement of SNS. The most threatening factor for small organizations was competitors.

On the other hand, medium sized businesses had some different barriers. One of them was control of negative comments arising from SNS. Another barrier was the fewer responses at first when setting up SNS with the Return on Investment (ROI) of SNS being hard to measure. Finally, the barriers and challenges related to large enterprises predominantly surrounded “keeping the company’s image”.

Discussion

The general analysis reflects similar trends among small, medium and large organizations as most of the companies use SNS as a marketing tool which is consistent with the results of a study by Eileen and Rebecca (2011) who found that the majority of businesses employ SNS as a marketing tool.

Gillin and Schwartzman (2011) stated that many organizations create a Facebook and Twitter account because they have the biggest populations, which reflects our finding.

Additionally, Derham et al. (2011) stated that the business cannot be sustained with the traditional marketing tools. Our study shows that most of the current companies that were interviewed are using both the traditional marketing tool as well as SNS.

A study by Hanna et al. (2012) indicated that a significant challenge for both large enterprises and SMEs is the lack of knowledge in the measurement and the effective use of SNS. Our results support and augment these findings by showing that SMEs are uncertain about measuring effectiveness.

DiStaso et al. (2011) indicated that another challenge facing organizations is the lack of expectation of what might be written (comments, complaints, negative posts).

Another finding is that large enterprises are concerned about their brand. In contrast, SMEs wanted to have a presence in the market. Their aim is to get more customers and gain benefit from customer.

Finally, the expectation of this study was that SMEs use SNS more than their large counterparts. However, after conducting the questionnaire, there was little evidence to suggest that notion. Unlike previous studies, in our study we consider small, medium and large enterprises separately. Thus, our chapter uncovers the more intricate differences between small, medium and large organizations in terms of their use of SNS.

Conclusion

This section draws conclusion from our findings, recommending a stage model for cultivating a fruitful social networking site. It also highlights areas for future research.

The purpose of our research was to find whether there is a difference between SMEs and large enterprises in their usage, objectives, management, benefit, barriers of usage of SNS and to determine whether SMEs use SNS more effectively than large enterprises.

In general, small enterprises should use SNS to get more customers whereas medium enterprises need to build presence through several channels. Large firms should aim to build awareness and increase loyalty. Additionally, the study shows that the time allocated for employees who manage the SNS increases from small, medium and large enterprises.

Another notable result is that small enterprises recognize that having a strategy is important to success with a clear plan on how, why and when will the SNS be used for the benefit of the business.

The study should be interpreted in terms of a number of limitations. The sample size was small and therefore, further studies with larger samples are necessary. A larger sample may facilitate cross-industry comparisons. The study's findings indicated some industry types only begun to use SNS recently compared to others, so future research should focus on the impact of industry type on how quickly enterprises adopt SNS and how can we utilize these SNS to suite all types of business.

Furthermore, a longitudinal study may also uncover the link between SNS use and business performance to determine the most appropriate use through time. Our chapter highlights that limited tools are available to evaluate the effectiveness of SNS. Therefore, an investigation of the best tools to measure effectiveness for SMEs would also provide valuable research in the future.

Recommendations

Despite these limitations, this study attempted to incorporate both qualitative and quantitative findings and it is an important step in comparing the use of SNS by SMEs and large organizations. This recommendation outlines the options that organizations can consider in the adoption of SNS. Based on the results described, it is possible to arrive at different views of using social media in organizations as a stage model for cultivating a fruitful social networking site as described below

The Stage Model for Cultivating a Fruitful Social Networking Site

The rationale behind the stage model of fostering a fruitful social networking site is based on the analogy of a garden where everyone can come and grow plants of their choice. The garden can be small or medium or large. We assume that the traditional way of marketing is a walled garden which is expensive and highly crafted whereas, social media is about gardening in a more informal manner.

- Level 1: Choose your audiences (Finding the right seed in the right place)
- Level 2: Plan to use SNS (Testing the soil)
- Level 3: Use the right tools for your company (Digging)
- Level 4: Build the content of SNS (fertilizing, seeding)
- Level 5: Manage SNS (Gardeners)
- Level 6: Update the content (Watering)
- Level 7: Measurement "Like, retweet...etc" (Harvesting)
- Level 8: Manage complains and negative comment (Weeding, pruning)

REFERENCES

Berthon, P., Pitt, L., McCarthy, I., & Kates, S. (2007). When customers get clever: Managerial approaches to dealing with creative consumers. *Business Horizons*, 50(1), 39- 47.

Blaxter, L., Hughes, C. & Tight, M., (2001). *How to Research,(2nd ed)*.Buckingham: Open University Press.

Breslauer, B., & Smith, T. (2009). Social media trends around the world! The global web index (GWI). *ESOMAR Research*, Online Research, Chicago.

Buehrer, E., Senecal, S., & Bolman , E.(2005). Sales force technology usage—Reasons, barriers, and support: An exploratory investigation. *Industrial Marketing Management*, 34, 389–398.

Derham, R., Cragg, p.&Morrish,S. (2011). Creating Value: An SME And Social Media. *Association for Information Systems*, AIS Electronic Library (AISeL).

DiStaso,W., McCorkindale, T., & Wright, D. K. (2011). How public relations executives perceive and measure the impact of social media in their organizations. *Public Relations Review*, 37(3), 325-328. doi: 10.1016/j.pubrev.2011.06.005

e-Marketer. (2010). Leveraging best practices for socialmedia. Retrieved from http://www.emarketer.com/Article.aspx?R=1008057 Last accessed January 14, 2014

eBizMBA. (2012). Top 15 Most Popular Social Networking Sites. Retrieved from http://www.ebizmba.com/articles/social-networking-websites, and http://www.compete.com/us/ Last accessed January 14, 2014

Eileen ,F.&Rebecca,R. (2011). Social interaction via new social media: (How) can interactions on Twitter affect effectual thinking and behavior?. *Journal of Business Venturing*, 26 (1), 1-18.

Geho, P.& Dangelo,J. (2012). The Evolution of Social Media As a Marketing tool for Entrepreneurs. *Entrepreneurial Executive*, 17, 61-86.

Gillin, P., &Schwartzman, E. (2011). Social Marketing to the business Customer: Listen to your B2B market, generate major account leads, and build client relationships.*Canada John Wiley& Sons, Inc.*,Hoboken, New Jersey.

Hanna, R. et al. (2012). We're all connected: The power of the social media ecosystem. *Business Horizons*, 54(3), 265-273.

Hoffman, L. & Fodor, M. (2010). Can You Measure the ROI of Your Social Media Marketing?.*MIT Sloan, Management Review*, 52(1).

Hurd, B. (2012). 10 things your social media policy should cover. Retrieved from http://123socialmedia.com/10-things-your-social-media-policy-should-cover/ Last accessed January 14, 2014

Kaplan,M. & Haenlein, M. (2010). Users of the world, unite! The challenges and opportunities of Social Media. *Business horizons*, 53(1), 59-68.

Kietzmann, j. , Hermkens, K., McCarthy,L.&Silvestre, B. (2011). Social media? Get serious! Understanding the functional building blocks of social media. *Business Horizons*, 54, 241- 251.

Kozinets,V. (2002). The field behind the screen: Using netnography for marketing research in online communities. *Journal of Marketing Research*, 39(1), 61-72.

Mangiuc, M. (2009). Measuring Web 2.0 Efficiency. Annales Universitatis Apulensis Series Oeconomica, 1(11)

Muniz, M., & O'Guinn, C. (2001). Brand community. *Journal of Consumer Research,* 27(4), 412-432.

Nielsen-Wire. (2010). Led by Facebook, Twitter, Global Time Spent on Social Media Sites up 82% Year over Year. Retrieved from http://blog.nielsen.com/nielsenwire/global/ledby-facebook-twitter-global-time-spent-on-social-media-sites-up-82-year-overyear/ Last accessed January 14, 2014

Policy, Social Media. (2011). Retrieved from http://123socialmedia.com/social-media-policy/ Last accessed January 14, 2014

Qurashi, A. (2009). The Power of Social Networking Sites in the context of Online Recruitment. (BA Honours), Robert Gordon University, United Kingdom.

Robson, C. (2002). Real World Research, (2nd ed). Oxford: Blackwell.

Saunders, M., Thornhill, A. & Lewis, P. (2009). Research methods for business students, (5th ed). Essex, England.

Venkatesh, V., & Davis, D. (2000). A theoretical extension of the technology acceptance model: Four longitudinal field studies. *Management Science*, 46(2), 186–204.

Waters, R. D., Burnett, E., Lamm, A. & Lucas, J. (2009). Engaging stakeholders through social networking: How nonprofit organizations are using Facebook. *Public Relations Review*, 35(2), 102-106.

Yin, R. (2009). Case Study Research: Design and Methods, (4th ed). Thousand Oaks, CA: Sage Publications, Inc.

In: Technology Innovation Leadership ...
Editors: G. Rampersad and F. Patel

ISBN: 978-1-63117-388-2

Chapter 6

ADOPTION OF e-COMMERCE SERVICES IN THE KINGDOM OF SAUDI ARABIA

Abdullah Saleh Alqahtani*[*], Robert Goodwin and Denise de Vries
Flinders University, Australia

ABSTRACT

In the Kingdom of Saudi Arabia, western and eastern economies have merged, resulting in an interesting economic environment. Within this environment, e-commerce has developed, increasing the potential of the local economy.

Despite the significant resources of Saudi Arabia's oil-based economy, the government has directed its income towards developing the non-oil sectors of the economy. This chapter examines the cultural, economic, and social factors that have contributed to the development of e-commerce in Saudi Arabia. Additionally, the chapter presents some of the obstacles and challenges surrounding e-commerce in Saudi Arabia, within the economy's current context.

---

[*] shamer_84@hotmail.com.

INTRODUCTION

As the birthplace of Islam, Saudi Arabia contains two of the religion's holiest places: Mecca and Medina. Even though it is a deeply religious country, Saudi Arabia has adopted many modern practices, from culture to commerce and technology. The official title of the King of Saudi Arabia is the "Custodian of the Two Holy Mosques." Abd Al-Aziz bin Abd al Rahman Al Saud (Ibn Saud), the previous King of Saudi Arabia, achieved the unification of almost the entire Arabian Peninsula and formed the foundation for the modern state. This feat took almost 30 years, finally being realized in 1932 (Central Intelligence Agency [CIA], 2009).

Government control in Saudi Arabia is intense, yet the country's development and economic growth place it among the richest countries in the world. Oil-sector revenue accounts for almost half (45 percent) of GDP and 90 percent of exports (CIA, 2009).

Although globalization is perceived in many countries as a way of replacing local concepts and values with western/American concepts and values, the Saudi Arabian perspective on and approach to globalization and its implications are different. The Kingdom of Saudi Arabia (KSA) has refused the replacement of its local values, culture, and policies; instead, Saudi Arabian governments have taken parts of western globalization concepts and adapted them to match their local cultural and economic characteristics. As a result, the development of the local economy may have been delayed.

Globalization has enabled countries like Saudi Arabia to develop and grow, but it has also created some limitations and challenges. E-commerce may have been less successful in Saudi Arabia than in many other countries not only because of cultural factors, but also because of setup costs and resistance to change. The following chapter will discuss these aspects.

E-COMMERCE ADOPTION AND THE CURRENT LEVEL OF E-COMMERCE USAGE IN KSA

The general popularity of the Internet has strengthened the growth of e-commerce, which is a global phenomenon in both developed and developing countries. Several studies have been conducted in developed countries to measure consumers' e-commerce adoption levels. However, there is a need to measure e-commerce adoption levels in other countries generally and in developing countries in particular.

Many factors influence the adoption level of e-commerce. Among the most obvious are trust in online sellers, national cultures, e-commerce infrastructure, and consumers' educational levels. E-commerce's high levels of uncertainty and risk—as the process is virtual and lacks control mechanisms—means that trust is a vital factor in its use. Buyers develop trust in processes based on their national culture. The tendency to take risks and address uncertainty is also influenced by national culture (Hofstede, 1980). Additionally, infrastructure and education can be considered the backbone of e-commerce; they complement each other because without them, e-commerce activities cannot thrive. Thus, it is obvious that poor infrastructure and lack of sufficient knowledge hinders online purchasing (Kashif, Hussain, & Umer, 2011).

E-COMMERCE IN DEVELOPING COUNTRIES

E-commerce has been relatively successful in developed economies. Studies have revealed that, in recent years, e-commerce has also become popular in developing countries (Molla & Licker, 2005; Scupola, 2009). However, other studies have revealed that e-commerce adoption has fallen below expectations in developing countries (Alam, Khatibi, Ahmad, & Ismail, 2008; Bélanger & Carter, 2008; Khalifa & Shen, 2008; Nasco, Toledo, & Mykytyn, 2008; Sulaiman & Abdullah, 2011). Developing countries face various barriers and issues related to e-commerce adoption. These are discussed below.

ISSUES IN ADOPTING E-COMMERCE

Unlike developed countries, developing countries lack the necessary financial, legal, and physical infrastructure for the growth of e-commerce. Therefore, e-commerce adoption in developing countries differs radically from adoption in developed countries (Mohamad & Ismail, 2009). Primarily, practices and cultures in developing countries limit the applicability and relevance of e-commerce models utilized by developing countries. For example, businesses in developing countries are generally smaller and less complex than in developed countries. Although it may seem that e-commerce adoption would be straightforward for a smaller, less complex businesses, the lack of sufficient resources to invest in e-commerce is a hindrance (Al-Otaibi, Al-Zahrani, & Rasheed, 2009; Ha & Stoel, 2009; Sulaiman & Abdullah, 2011).

WHY SELLERS ARE RELUCTANT TO USE E-COMMERCE?

Many factors influence a seller's decision to use e-commerce. These factors are explained by Abdalla, Ball, Berger, & Cleary (2008), and are summarized as follows.

Establishing an e-Commerce Department

It is necessary for the government to establish a department or ministry that is accountable for e-commerce activities and regulation. In order to gain the trust of the people and to be integrated with other government services, this department should work closely with other relevant departments/ministries (Abdalla et al., 2008).

Cultural and Religious Values

As discussed earlier, cultural and religious values affect levels of e-commerce adoption. E-commerce adoption must not conflict with the country's culture and values (Abdalla et al., 2008). People may be reluctant to use e-commerce if it has an impact on their culture or religion. Therefore, businesses using e-commerce must ensure that it does not affect the religion or culture of the country in which they are operating. For instance, selling alcohol online in Muslim countries may drive shoppers away from the business. In the worst case, many consumers might develop a negative perception of e-commerce on the whole based on that particular business. In the future, this perception is likely to affect users' motivation to shop online (Hernandez, Jimenez, & Martín, 2009).

Use of Local Language

Language is one of the major barriers to the successful adoption of e-commerce in developing countries. Many inhabitants do not speak foreign languages. If websites are available in only one language and it is not understood by consumers, poor adoption will result (Abdalla et al., 2008). Therefore, e-commerce businesses need to use local languages to encourage consumers to use the Internet and shop online. In this way, consumers will recognize that their language, culture, and traditions remain safe and respected.

Introducing New Strategies

E-commerce focuses not only on webpages for businesses; it is an entirely new way of doing business. New strategies are required at various organizational levels. In developing countries, many businesses grapple with the challenge of new e-commerce strategies. The government and/or related departments should help businesses to design appropriate e-commerce strategies (Abdalla et al., 2008).

POTENTIAL OF THE SAUDI ARABIAN MARKET

According to 2004 statistics, Saudi Arabia has 23 million dwellers and a US$6.3 billion information technology (IT) product market (including computer sales of $850 million). In contrast, the growth potential of e-commerce is not very encouraging because of the lack of appropriate IT regulations in the country, the slow penetration of the Internet, and the lack of advanced Internet technologies. In 2001, Saudi Arabia was one of the largest IT markets among Arab countries because of its record sale of IT products (Saleh, 2008). Although this growth appears impressive in terms of volume, it is not as encouraging when compared to other countries in the Middle East (Al-Hoymany, 2002). Currently, the government of KSA is planning to convert Riyadh (the capital city) into a hub of IT telecommunications, with an investment of $7 billion. This implies that KSA might be behind its smaller Middle Eastern neighbors with respect to Internet speed and related technologies, although the government is investing heavily to overcome this gap (Ministry of Communications and Information Technology, 2004; Saleh & Snow, 2004).

E-COMMERCE INFRASTRUCTURE IN KSA

In Saudi Arabia, the three most important components of e-commerce are as follows: a) telecommunication and Internet networks, b) payment mechanisms, and c) delivery of orders. Local authorities have begun procedures for implementing proper e-commerce regulations, including the creation of e-government facilities. Online business has revealed its benefits and companies have begun to increase their use of such facilities. Large businesses have implemented e-commerce infrastructure to increase and stabilize their activity (Esty, Levy, & Winston, 2003). The three components of e-commerce will be discussed in the following sections.

Telecommunication and Internet

The telecommunication market in the Middle East has developed significantly in the last few years, with Saudi Arabia experiencing rapid market increases. With a total e-commerce expenditure of US$4.7 billion at the end of 2005, Saudi Arabia is expected to reach US$60 billion by 2020. The telecommunications department of KSA aimed to increase the availability rate of telephone lines to 25 lines per 100 people by the end of 2005. This represented a huge increase from 10 lines per 100 people in 1996 (Esty et al., 2003), but the aim was achieved. The Saudi government also approved a plan to invest more than US$10 billion to improve the information and communications technology (ICT) infrastructure of the country between 2005 and 2020 (Esty et al., 2003). Due to rapid ICT developments in KSA, there is currently a high demand for workers with skills in network management, data warehousing, and web administration (US–Saudi Arabian Business Council, 2005).

Payment Mechanisms

The ease of online payment mechanisms plays an imperative role in the development of e-commerce in any country. In 2004, SambaConnect was introduced as the first mechanism for online payment in KSA (Saleh, 2008; Samba Financial Group, 2013). In early 2005, tremendous improvements were made in KSA with respect to online payment, with the first-quarter volume of online payments recorded as $100,000 (VISA-Europe, 2013).

It is expected that e-commerce will flourish in KSA in the coming years. It is also anticipated that recent security measures (like 'Verified by Visa') and greater acceptability of online payments (for example, Arabian Airlines now accepts SambaConnect) will increase public confidence in e-commerce (AME-Info, 2005). Bélanger & Carter's (2008) study showed that customer trust in online payments is extremely important for the successful adoption of e-commerce.

Means of Delivery

The smooth and fast delivery of online orders is a key factor for the success of e-commerce in any part of the world. In KSA, the Saudi Post is responsible for the delivery of mail items. In 2005, the Saudi Post approved a project to deliver 5 million mail items directly to customers' doorsteps instead of through private post-office boxes. This project was successful and Saudi Post gained valuable experience, helping it to plan and implement such a system on a larger scale.

After the successful implementation of this system, e-commerce sellers should be able to deliver their orders to buyers directly, easily, and quickly. These and similar measures should help to establish an address verification system (AVS) such as that already implemented in the United States and Europe. AVS ensures that the billing address of the card being used for payment matches the address provided by the customer (Saleh, 2008).

E-Commerce and Basic Economic Elements in KSA

Saudi Arabia relies on oil as the main sector of its economy. Having the largest oil reserve and the most significant production in the Organization of the Petroleum Exporting Countries (OPEC), Saudi Arabia provides a third of the oil output on the market. The capacity of Saudi reserves is almost 10 million barrels per day, allowing the country to control the oil prices on the market. Although the Saudi economy is oil-based, the government is attempting to diversify and has developed sectors such as steel, fertilizers, and petrochemicals, without ignoring two of its traditional income sources: agriculture and fishing (Swiss Business Hub GCC, 2010).

Oil revenues have allowed the Saudi economy to develop at an accelerated rate. In 2007, the nominal GDP increased by 12.4 percent (to $346.9 billion), and the real GDP increased by 4.2 percent (to $213.3 billion). The government is careful about investing the surplus of its economic growth and is attempting to find an efficient balance between the oil and non-oil sectors of its economy. In 2007, the Saudi government invested 16 percent more in the oil sector and 10 percent more in the non-oil sectors. The government has designed a strategy to allow the economy to develop its non-oil sectors and to use this diversification to develop the employment area. The macroeconomic policies and the core structure of the Saudi economy are focused around important projects with long-term impacts on the economy (Swiss Business Hub GCC, 2010).

Foreign investment is developing in Saudi Arabia and companies like the Saudi Basic Industries Corporation (SABIC) and ExxonMobil have increased their interest in the petrochemical sector. Foreign companies will cooperate with regions like Jubail and Yanbu because of the incentives given to foreign investors. Jubail city, which is relatively large with over 90,000 residents, supports numerous industrial plants, institutes, and colleges. The city has 40 industrial plants and almost 100 light industrial plants. With a little over 70,000 residents, Yanbu city allows major industrial activities along the Red Sea Coast, yet facilitates the export of local manufactured goods. Even though Yanbu city has almost the same number of residents as Jubail city, it has just a little over half the number of basic and secondary plants, and only a third the number of supporting light industrial plants (Swiss Business Hub GCC, 2010).

E-Commerce Services Problems in KSA

The growth of e-commerce is not adequate (Albadr, 2003; Aladwani, 2003; Communications and Information Technology Commission [CITC], 2007; Agamdi, 2008), despite the fact that Saudi Arabia's ICT market demonstrates the largest and fastest growth in the Middle East (Al-Otaibi & Al-Zahrani, 2003; US Department of Commerce, 2008; Alfuraih, 2008; Saleh, 2008). Among large companies in KSA, B2B is the most frequently used mode of e-commerce in Saudi Arabia (Rayed et al., 2011).

Many researchers have tried to elicit the reasons for the slow evolution of e-commerce in the Middle East in general, and Saudi Arabia in particular. The reasons ascertained include sub-standard ICT infrastructure, trust and privacy issues between buyers and sellers, and the lack of related regulations and legislation that protect both buyers and sellers (Aladwani, 2003; Al-Solbi & Mayhew, 2005; Alfuraih, 2008; Agamdi, 2008; Alrawi & Sabry, 2009; Alghaith, Sanzogni, & Sandhu, 2010).

Cultural Influence and Online Buying/Selling

A country's culture plays a major role in the success or failure of e-commerce. This is especially pertinent in Saudi Arabia, where people like to physically visit retailers to shop. Most inhabitants of KSA are emphatic about their preference for retail shopping over e-commerce and this preference is considered the key reason for the country's relatively slow progress in e-commerce (Rayed et al., 2011).

Retailers in KSA share a belief that e-commerce in not a strong movement in the country. In the western world, going online to check prices and make purchases has become a part of life. In contrast, such behavior is unusual in KSA. While the number of persons who use the Internet is high, the number of persons who make online purchases is low (Rayed et al., 2011). Such a consumer culture does not encourage retailers to embrace e-commerce (Rayed et al., 2011).

Overall, the custom in KSA is for consumers to inspect goods physically before buying. If this trend continues, there will be little room for e-commerce. Customs such as this are difficult and slow to change. For this reason, retailers do not deem e-commerce to be a viable option in KSA (Rayed et al., 2011).

The non-enthusiastic behavior of KSA inhabitants toward e-commerce is further confirmed by those retailers currently offering e-commerce services. There is consensus among these retailers that online sales cannot be compared with a physical retail shop and that online business is not a worthwhile investment because of consumers' habits. This situation contrasts with that in Europe and America, where consumers are increasingly buying online (Rayed et al., 2011). Interestingly, KSA retailers face a low response rate from buyers even when they offer online discounts and free delivery. An online seller in Rayed et al. (2011) study received only 3–5 online orders per week, which was too few compared to its retail shop.

Many factors influence the willingness of KSA inhabitants to adopt e-commerce. These include the strict online payment mechanisms (i.e., credit cards) and the delivery infrastructure in KSA. Despite these problems, some retailers are strongly committed to e-commerce. Moreover, research has showed that there is a specific group of KSA consumers who buy online; members of this group have studied overseas in western countries where they have been exposed to e-commerce (Rayed et al., 2011).

Competitive Advantage of Online Business

Business in open world markets is highly competitive and gaining competitive advantage is the key to success. The same is true for e-commerce. Some KSA retailers have already expressed concern that they cannot achieve competitive advantage in e-commerce. The main issue faced by retailers regarding e-commerce is delivery fees, which boost the total cost of transactions. Clearly, if customers are forced to pay a higher price for e-commerce transactions, they are not going to buy online. According to computer accessories retailers in KSA, computer shops are widespread in Jeddah, offering comparative prices, and it is simply not possible for an online retailer to compete with these prices due to postage and insurance costs (Rayed et al., 2011). This is further illustrated by a leading manufacturer of Islamic watches and clocks. This manufacturer only sells online internationally because similar watches and clocks are easily available in KSA and it is not possible to sell these online there due to the additional price of postage (Rayed et al., 2011).

Delivery Issues

The parcel-delivery system is fundamental to the success of e-commerce. KSA retailers use many methods to deliver goods purchased online. Some companies operate their own delivery systems, whereby a customer places their order online or by phone and the goods are delivered to the company's local office to be collected by the customer. Home delivery can also be arranged but incurs an additional fee, and the customer has to explain their location to the driver or have a live call with the driver during delivery (Rayed et al., 2011). In this mode of delivery, the customer's mobile phone number is the key. If the number is incorrect, unattended, or unresponsive, the customer and the seller both face problems. Additionally, the addresses of residences are not always well known. All these issues are currently handled using the mobile phone number and the customer may be called many times during this process.

Some customers are also concerned about the safe delivery of their order if it includes fragile items (Rayed et al., 2011).

Few Online Payment Options

Buying and selling (either online or at a physical outlet) relies on payment for the good. In a physical outlet, the customer can see and then pay for the good. In e-commerce, trust must be established between customer and seller. Most online businesses receive payment beforehand, but the customer also requires assurance that the promised good will be delivered in appropriate condition. Human perception plays an important role in trust. For example, customers are comfortable providing credit card details to trusted companies like Amazon.com but may be hesitant to provide details to newly established companies. The same is true for online companies in KSA; consumers are doubtful about the privacy of their financial information as well as the quality of good received (Rayed et al., 2011).

Customers' lack of trust means that sellers are also hesitant to invest in online payment mechanisms. Some retailers show prices on their websites but ask customers to use bank transfers to place their orders. Customers tend to be more confident about this payment mechanism even though it is slow and time consuming.

Poor ICT Infrastructure

The "e" in e-commerce is all about technology and technological infrastructure. This ICT infrastructure plays a key role in the development of e-commerce. Although there is a strong belief in KSA that consumers will eventually turn to e-commerce, ICT infrastructure needs to be appropriate before this can occur. For example, Wi-Fi needs to be widely available and more people need to have access to broadband, especially in remote areas. Moreover, safe payment mechanisms must be in place. Additionally, reliable and efficient telecommunication with an order delivery network throughout the country are essential. Furthermore, it is necessary to have an accessible telecommunication network that customers can access easily to make purchases. Moreover, customers expect fast delivery of their purchased items, which requires a good delivery network. Non-reliability of either of these two networks (telecommunication and delivery) can delay the delivery of orders.

Finally, an integrated solution—like that implemented in the stock market in KSA—is the best option for the successful integration of e-commerce (Rayed et al., 2011).

Lack of e-Commerce Legislation

During the execution of any business transaction, the related regulations and rules are vital. These legislative regulations and rules cover the rights and duties of both sellers and buyers, and the procedures in case of any dispute. Currently, there are no clear laws in place in KSA regarding e-commerce. Both buyers and sellers need to be sure that there are rules in place to protect them. This will help to develop trust—an important ingredient for the success of e-commerce—between buyers and sellers. Legislation should be clear policy from the government, developed through consultation with all parties concerned.

Lack of Public Trust

Some buyers and sellers do not trust online systems. This concern is associated with online payments. In KSA, there are many stories of stealing/hacking of credit card numbers on the Internet (Rayed et al., 2011).

Suitability of Products for Online Selling

Many factors influence the decision of businesses to offer specific types of products. These factors include culture, environment, and social values. In a simple example, the sale of some types of fragile goods is not possible if special delivery arrangements cannot be made. Another example is that in KSA, where temperatures can reach up to 50°C, the sale and delivery of food items require special arrangements. In other scenarios, buyers are hesitant to buy certain items online for fear that items will not be delivered in good condition. In 2009, a chocolate manufacturing company conducted research and found that a considerable portion of online buyers are hesitant to buy such heat-sensitive products online, preferring to buy them from a physical shop (Rayed et al., 2011). Similarly, it is difficult for a beauty company to sell perfumes, makeup, cosmetics, and body lotions because buyers want to try out these items (feel, smell, and so on) to check that they are to their taste before they buy. Additionally, online sale of electric and electronic items (fridges, air conditioners, televisions, and so on) has been slow because these can not only malfunction in high temperatures but also incur considerable delivery charges, making the total transaction price much higher than that paid at physical outlets.

In contrast, for those products that are not widespread in the market, the idea of online selling can be appealing; the lack of availability of these items in the market provides a strong competitive advantage to online sellers.

Lack of e-Commerce Experience

Many businesses in KSA could successfully adopt e-commerce but do not do so because of a lack of knowledge/experience. Moreover, some businesses feel that e-commerce is troublesome, requiring substantial investment of both time and finances.

Such ignorance plays an important role in the slow growth of e-commerce because sellers exchange opinions with each other, building a consensus among the business community regarding e-commerce.

Low Profit Margins in Online Business

The main motive of any business is to earn a profit. In simple terms, if an idea is not profitable, it is rejected. This ideology applies to e-commerce. Whatever the reason, e-commerce in KSA will only flourish when it is profitable. Many factors contribute to the relatively high costs of e-commerce in KSA, including delivery charges and payment gateways fees.

Resistance to Change

Resisting change is a natural human tendency because familiarity only develops over time. For this reason, KSA retailers prefer the traditional (physical) way of selling. Their preference is underscored by the preference of customers, who like to inspect products physically before purchasing.

High Setup Costs for Online Business

For some KSA retailers, the initial setup cost for e-commerce is another major hurdle. Starting up an e-commerce business involves numerous outlays, including the design and development of a commercial website, online payment-handling fees, training of employees, and delivery of goods. In developed countries, there are many ways to cope with these initial setup costs—for example, governments may offer tax rebates as an incentive to new online retailers. These types of incentives do not exist in KSA.

Failure of the Postal System

As explained previously, many KSA buyers do not buy online because they do not have a home mailbox. Although there are few opportunities to buy online within KSA, KSA citizens can easily buy from international online websites, but most do not because of the home address difficulty.

Mr. Muhammad (a customer) explained that when he needs to tell someone where he lives, he cannot provide a direct address with a street/house number and so forth. Currently, if he wants to explain to someone where he lives, he can only provide the name of the suburb before trying to explain the area and the nearest streets. This is problematic for both buyer and seller, especially when the situation is replicated numerous times (Al-Harby, 2006; Rayed, Steve, & Osama, 2011; Tokuro, 2003). Those KSA buyers who buy online from international websites have private mailboxes with courier companies or receive the order in the mailbox of a relative or friend.

Internet Censorship

The Saudi government first allowed the public to use the Internet in 1997 and since then, it has made a strong impact on the Saudi lifestyle. It has become an integral part of Saudi society in the areas of education and the economy. Like any technology, the Internet has its pros and cons. No doubt, the Internet can change lifestyle in a beneficial way but, recently, it has also emerged as a tool for destabilizing the political situation. For example, during Iran's 2009 presidential election, Internet-based social media and blogs were used to organize political protests. Therefore, the Saudi government is using its entire means to sensor the Internet to protect itself and its people from its harmful use. These means include the blocking of websites, information filtration, and suppression. Such means fall under the broader term of 'Internet censorship' (Khalid, 2010).

Internet censorship in KSA is similar to the types of censorship already in place in television, radio and print media. However, publishing and spreading information on the Internet is easier and faster than using these other means of communication. Therefore, the Saudi government has created a new department called the Communications and Information Technology Commission (CITC), which is primarily responsible for filtering illegal Internet content and monitoring Internet service providers (ISPs) in KSA. This strict Internet censorship is criticized by international organizations that support freedom of speech, but the Saudi government has realistic reasons to support its censorship activities (Khalid, 2010).

The first reason that the Saudi government presents in favor of Internet censorship is the increasing availability of pornographic content on the Internet. It is estimated that 4.2 million websites contain pornographic material, and pornography is abhorred in KSA as a result of its culture and strong religious values (Khalid, 2010).

Another leading reason for Internet censorship is to ensure that national security is not threatened in any way. Currently, one of the major threats is terrorism. Many terror organizations run their websites to convey inappropriate material and also as a means of communication (i.e., to establish contact with their associates). The Saudi government argues that for this reason, Internet censorship is necessary. Moreover, in KSA there is little government opposition and the government restricts access to any type of website that spreads opposing political thought (Khalid, 2010).

Unfortunately, these measures may limit the freedom of speech in KSA. For example, blogs demonstrate an accessibility problem due to Internet censorship. Moreover, the government arrests those who criticize the government or royal family. Khalid (2010) notes that there is a negative effect of Internet censorship as it kills freedom in the country. Although CITC has introduced a service on its website through which the public can request the blocking/unblocking of websites, there is no timeline or guarantee that public requests will be entertained. Moreover, if a legitimate e-commerce business is blocked by mistake, no compensation is given (Khalid, 2010).

The main reasons for Internet censorship in KSA are to restrict pornographic content, to improve national security, and to restrict political ideas that oppose the government. Khalid (2010) argues that the misuse of

Internet censorship can slow the (already very limited) pace of the integration of e-commerce in KSA.

E-Commerce and Strategies of Developing Countries

The reduced cost of e-commerce information search and transactions represents a significant advantage in developing countries. For example, the low costs allow companies to improve the efficiency of their operations, which consequently reduces the processing time for payments or credit verifications. According to research, customers and markets, process technology, product design, as well as source financing, are significant elements that must be considered in e-commerce.

E-Commerce's Impact on International Trade and Employment

The impact of e-commerce on international trade and employment is apparent. In the short term, more benefits exist for developed countries, but in the long term, developing countries stand to gain greater benefits from e-commerce. International trade is positively influenced by e-commerce because it allows developing countries to absorb information, products, and services without the restrictions that exist in traditional commercial relationships. Moreover, countries that are open to imports from developed countries are likely to benefit from the latest technology and knowledge. In conclusion, the impact of e-commerce is both positive and negative; it will create jobs in new and developing fields while eliminating other jobs that involve human input or other human activities that support classic commerce (Nuray, 2011).

E-Commerce's Impact on the Local Retail Market

A major body of research has explored how e-commerce has helped to reduce prices (Maris, Ali, Chad, & Onsel, 2008); however, little work has been done to investigate the types of products suitable for e-commerce. This suitability depends on many factors, including cultural values, economic conditions, and the country's political situation. Research has investigated how such factors allow some industries to attract clients to the detriment of other industries. For example, in the US, customers are paying less for travel, brick-and-mortar bookstores have lost customers to online stores, and auto-dealers have lost ground to websites that allow customers to obtain information from the comfort of their homes. Even though all three industries have recorded significant changes, the mechanism through which e-commerce has determined consumer and company behavior is different in each case (Maris et al., 2008). Bookstores and auto-dealers have lost out to e-commerce businesses because of local factors such as the market exit of stores and industry-wide declines, while travel agencies have lost ground to online agencies because of airlines' reduced commissions, which are the result of customers' increased usage of online ticket-purchasing facilities (Maris et al., 2008; Carter & Weerakkody, 2008).

GAPS IN EXISTING LITERATURE ON E-COMMERCE IN KSA

The following key areas are specifically related to e-commerce in Saudi Arabia:

Advertising and Marketing Barriers

Like every society, Saudi society has its own traditions and culture. For example, KSA—unlike many other countries—prefers not to use women to advertise or market products. As a result, in order to boost e-commerce, alternatives need to be found to advertise and market e-commerce products in KSA. Although similar studies have been performed (Pires & Aisbett, 2003; Al-Qirim, 2008; Carter & Weerakkody, 2008), there is a need for in-depth studies of e-commerce adoption with respect to KSA.

Language, Culture and Religious Limitations (Halal Food)

Saudi Arabia's rich Arab culture has been developed over centuries. Its language, culture and religious values make it distinctive from other non-Arab regions. For example, a food e-commerce business in Saudi Arabia must offer Halal food simply to operate.

Media Selection

Electronic media is now intrinsic in any marketing and advertising campaign. Considering the cultural and religious values of Saudi society, it is vital that any e-commerce business selects appropriate media (i.e., television, radio, newspapers, and/or online) in which to advertise.

Customer Service/Satisfaction and Relationships

In any online business, buyers and sellers might not meet with each other in person. This is a special feature of e-commerce that differs from traditional face-to-face buyer–seller transactions. This "non-eye-contact" mode of e-commerce requires excellent customer service/satisfaction. Due to the aforementioned current e-commerce problems in Saudi society, it is particularly important that any e-commerce business maintains a high standard of customer service/satisfaction and builds a relationship of trust.

CONCLUSION

The Saudi Arabian economy has developed considerably in the last few years as globalization has occurred. However, the process of e-commerce adoption has been difficult in KSA because of economic, cultural, and social factors. A lack of legislation and regulations to protect both buyers and sellers has created a lack of enthusiasm from clients and companies. Among the most significant factors contributing to the low e-commerce adoption levels are a resistance to change, limited delivery options and several other problems that have impeded the finalization of e-commerce. These other impediments include the costs of setting up an online business and the resistance to change related to some products such as heat-sensitive goods, which clients prefer to buy from traditional retail shops. These barriers are likely to reduce over time, as the economy adopts more western elements of development. Nevertheless, KSA needs to overcome certain challenges and obstacles in order to develop its e-commerce environment.

REFERENCES

Abdalla, H., Ball, D., Berger, H., & Cleary, P. (2008). The three-quarter moon: A new model for e-commerce adoption. *Communications of the IBIMA*, *4*, 88–96.

Agamdi, A. (2008). *E-commerce implementation challenges and success factors in the Kingdom of Saudi Arabia.* Paper presented at the Nineteenth National Computer Conference: The Digital Economy and ICT Industry, Riyadh, Saudi Arabia, 1–5.

Ahmad S. (2005). *The application of information systems in the Jordanian banking sector: A study of the appetence of the Internet* (Unpublished PhD Thesis), University of Wollongong, New South Wales.

Aladwani, A. (2003). Key Internet characteristics and e-commerce issues in Arab countries. *Information Technology & People*, *16*(1), 9–20.

Alam, S., Khatibi, A., Ahmad, M., & Ismail, H., (2008). Factors affecting e-commerce adoption in the electronic manufacturing companies in Malaysia. *International Journal of Commerce and Management, 17*(1/2), 125–39, Emerald Group Publishing Limited.

Albadr, B. (2003). E-commerce. *Science and Technology*, *65*, 14–19.

Alexandra, K., & Boudkov, O. (2012). *Report: E-commerce in Saudi Arabia: Driving the evolution, adaption and growth of e-commerce in the retail industry.* Retrieved from http://sachaorloff.wordpress.com/2012/06/16/e-commerce-in-saudi-arabia-driving-the-evolution-adaption-and-growth-of-e-commerce-in-the-retail-industry/

Alfuraih, S. (2008). *E-commerce and e-commerce fraud in Saudi Arabia: A case study*. Paper presented at the Second International Conference on Information Security and Assurance, Busan, Korea, 176–80.

Alghaith, W., Sanzogni, L., & Sandhu, K. (2010). Factors influencing the adoption and usage of online services in Saudi Arabia. *Electronic Journal of Information Systems in Developing Countries*, *40*(1), 1–32.

Al-Harby, F. (2006). *Barriers of e-commerce and e-government in Saudi Arabia* (Unpublished MSc. thesis), School of Computing, Informatics and Media, University of Bradford, England.

Al-Hoymany, F. (2002). *E-business and PKI in Saudi Arabia, 2002*. Retrieved from http://www.idsc.gov.eg/Conferences/SeminarEbusiness/E-business-in-Saudi.Pdf

Al-Otaibi, M., & Al-Zahrani, R. (2003). *E-commerce adoption in Saudi Arabia: An evaluation of commercial organizations' web sites*. Retrieved from http://faculty.ksu.edu.sa/DrRasheed/My%20Publications/Saudi_Websites.pdf

Al-Otaibi, M., Al-Zahrani, B., & Rasheed, M. (2009). *Electronic commerce in the Kingdom of Saudi Arabia.* Retrieved from http://repository.ksu.edu.sa/jspui/handle/123456789/3262

Al-Qirim, N. (2007). The adoption and diffusion of e-commerce in developing countries: The case of an NGO in Jordan. *Information Technology for Development, 13*(2), 107–31.

Al-Qirim, N. (2008). The adoption of e-commerce communications and applications technologies in small businesses in New Zealand. *Electronic Commerce Research and Applications, 6*(4), 462–73.

Alrawi, K., & Sabry, K. (2009). E-commerce evolution: A gulf region review. *International Journal of Business Information Systems, 4*(5), 509–26.

Al-Solbi, A., & Mayhew, P. (2005). Measuring e-readiness assessment in Saudi organizations: Preliminary results from a survey study. In I. Kushchu & M. H. Kuscu (Eds.), *From e-government to m-government (pp. 467–75).* Brighton, England: Mobile Government Consortium International LLC.

AME-Info. (2005). *E-commerce reaches new highs in Saudi Arabia*. Retrieved from http://www.ameinfo.com/61990.html

Bélanger, F., & Carter, L., (2008). Trust and risk in e-government adoption. *Journal of Strategic Information Systems, 17*(2), 165–76.

Carter, L., & Weerakkody, V. (2008). E-government adoption: A cultural comparison. *Information Systems Frontiers, 10*(3), 473–82.

Central Intelligence Agency (CIA). (2009). *The World Fact Book: Saudi Arabia*. Retrieved from https://www.cia.gov/library/publications/the-world-factbook/geos/sa.html

Communications and Information Technology Commission (CITC). (2007). *Internet usage study in the Kingdom of Saudi Arabia*. Retrieved from http://www.citc.gov.sa/English/Reportsandstudies/Studies/Documents/IT%20010%20E%20-%20BusinessEstablishmentReport2008.pdf

Datta, P. (2010). A preliminary study of ecommerce adoption in developing countries. *Information Systems Journal, 21*(1), 3–32.

Datta, P. (2011). A preliminary study of e-commerce adoption in developing countries, *Information Systems Journal, 21*(1). Retrieved from http://dx.doi.org/10.1111/j.1365-2575.2009.00344.x

Esty, D., Levy, M., & Winston, A. (2003). Environmental sustainability in the Arab world. *The Arab World Competitiveness Report 2002–2003*. New York, NY: Oxford University Press.

Ha, S., & Stoel, L. (2009). Consumer e-shopping acceptance: Antecedents in a technology acceptance model. *Journal of Business Research, 62*(5), 565–71.

Hernandez, B., Jimenez, J., & Martín, M. (2009). Adoption vs. acceptance of e-commerce: Two different decisions. *European Journal of Marketing, 43*(9/10), 1232–45.

Hofstede, G. (1980). *Culture's Consequences: International Differences in Work-Related Values.* Sage Publications, Beverly Hills.

Kashif, B., Hussain, R., & Umer, F. (2011). *E-commerce adoption: A comparative study of Sweden and Pakistan* (Unpublished master's thesis). Department of Management and Engineering (IEI), Linköping University, Sweden. (ISRN No. LIU-IEI-FIL-A--11-01020—SE).

Khalid A., (2010). *Internet Censorship in Saudi Arabia.* Retrieved from http://www.kalharbi.com/2010/05/Internet-censorship-in-saudi-arabia/

Khalifa, M., & Shen, K. (2008). Explaining the adoption of transactional B2C mobile commerce. *Journal of Enterprise Information Management, 21*(2), 110–24.

Maris, G., Ali, H., Chad, S., & Onsel, E. (2008). *E-commerce and the market structure of retail industries* (No. 14166). National Bureau of Economic Research (NBER) Working Paper. Retrieved from http://www.nber.org/papers/w14166

Ministry of Communications and Information Technology. (2004). *Information and Telecommunication Technology in Saudi Arabia.* Retrieved from http://www.itu.int/dms_pub/itus/md/03/wsispc3/c/S03-WSISPC3-C-0025!!MSW-E.doc

Mohamad, R., & Ismail, N. (2009). Electronic commerce adoption in SME: The trend of prior studies. *Journal of Internet Banking and Commerce, 14*(2), 1–16.

Molla, A., & Licker, P. (2005). E-commerce adoption in developing countries: A model and instrument. *Information & Management*, *42*, 877–99. Retrieved from http://dx.doi.org/10.1016/j.im.2004.09.002

Nasco, S., Toledo, E., & Mykytyn, P. (2008). Predicting electronic commerce adoption in Chilean SMEs. *Journal of Business Research, 61*(6), 697–705.

Nuray, T. (2011). The impact of e-commerce on international trade and employment. *Social and Behavioral Sciences*, *24*, 745–53.

Olatokun, W., & Kebonye, M. (2010). E-commerce technology adoption by SMEs in Botswana. *International Journal of Emerging Technologies and Society, 8*(1), 42–56.

Pires, G., & Aisbett, J. (2003). The relationship between technology adoption and strategy in business-to-business markets: The case of e-commerce. *Journal of Industrial Marketing Management, 32*(4), 291–300.

Rayed, A., Steve, D., & Osama, A. (2011). Issues influencing Saudi customers' decisions to purchase from online retailers in the KSA: A qualitative analysis. *European Journal of Scientific Research, 55*(4), 580–93.

Rayed, A., Steve D., & Waleed, A. (2011). Factors influencing e-commerce adoption by retailers in Saudi Arabia: A qualitative analysis. *Electronic Journal on Information Systems in Developing Countries, 47*(7), 1–23.

Rogers, E. (2003). *Diffusion of innovations* (5th edn.). New York, NY, Simon & Schuster.

Saleh, A. (2008). E-*commerce and e-commerce fraud in Saudi Arabia: A case study.* Paper presented at the IEEE International Conference on Information Security and Assurance, 176–80.

Saleh, A., & Snow, R. (2004). Taxonomy of e-commerce disputes. *WSEAS Transactions on Computers, 6*(3).

Samba Financial Group. (2013). *SambaConect*. Retrieved from http://www.sambaconnect.com/

Scupola, A. (2009). SMEs' e-commerce adoption: Perspectives from Denmark and Australia. *Emerald Group Publishing Limited, 22*(1/2), 152–166.

Sulaiman, A., & Abdullah, A. (2011). E-commerce adoption factors in Saudi Arabia. *International Journal of Business and Management 6*(9), 122–33.

Swiss Business Hub GCC. (2010). *Economic cities Saudi Arabia: Opportunity assessment for Cleantech companies*. Retrieved from http://www.osec.ch/en/filefield-private/files/6536/field_blog_public_ files/7968

Tokuro, M. (2003). A study on electronic commerce support systems based on users' preferences (Unpublished MSc. thesis), *The Japan Advanced Institute of Science and Technology*.

US Department of Commerce. (2008). *Doing business in Saudi Arabia: A country commercial guide for US companies*. Retrieved from http://export.gov/saudiarabia/doingbusinessinsaudiarabia/

US–Saudi Arabian Business Council. (2005). *The Telecommunications and Information Technology Sectors in the Kingdom of Saudi Arabia.*

Retrieved from http://www.us-saudibusiness.org/Telecom%20Report%20June%202005.pdf

VISA-Europe. (2013). Retrieved from www.visaeu.com

WikiBooks. (2013). *E-commerce in developing countries*. Retrieved from http://en.wikibooks.org/w/index.php?title=E-Commerce_and_E-Business/E-Commerce_in_Developing_Countries

In: Technology Innovation Leadership … ISBN: 978-1-63117-388-2
Editors: G. Rampersad and F. Patel

Chapter 7

THE INTERNET OF EVERYTHING: ASSESSING ITS IMPACT ON BUSINESS AND COMMUNITY

Giselle Rampersad*
Flinders University, Australia

ABSTRACT

The 'Internet of Everything' (IoE) whereby each human on average is currently connected to about 200 things (smart phones, tablets, sensors and systems) has had a radical impact on business but also on community for instance through the Arab Middle Eastern Spring. This chapter examines how countries can grapple with evaluating the influence of IoE to leverage its ability to improve its impact on their economies and communities in vital areas of education and health. In addition to its relevance to governments, businesses internationally are also interested in the IoE, for instance CISCO, Microsoft and IBM in pursuing business opportunities for developing new services to leverage this connectivity. However, the extent to which this increased connectivity presents opportunities or challenges for development is yet to be assessed. A significant factor for developing communities to partake in this connectivity is the underlying Internet infrastructure. Investing in Internet infrastructure is usually a major digital investment facing developing communities. Although billions of dollars are being spent internationally

* E-mail: giselle.rampersad@flinders.edu.au.

on such infrastructure, there are insufficient effective tools to monitor the impact of such investment. This chapter is significant as it provides a tool for assessing the contribution of this digital investment on the economy and community, and can be used by government agencies and businesses in developing communities to identify opportunities for interventions and applications as required.

INTRODUCTION

The Internet of Everything is growing in prominence and refers to the pervasive connectivity of things or objects including smart phones, sensors, systems and tablets (Atzori, Iera and Morabito, 2011). Estimating that IoE contributes $14.4 trillion in value, CISCO (2013) notes that during 2008, the number of things (devices including smart phones, tablets, sensors and systems) connected to the Internet exceeded the number of people on earth and predicts that by 2020, 50 billion devices will be connected to the Internet. Although billions of dollars are being spent on high-speed broadband Internet initiatives internationally, there are insufficient effective tools to monitor the impact of this connectivity. The World Economic Forum Global Information Technology Report (WEF 2011) has highlighted that the Finnish government has promised to provide high-speed broadband Internet (100 MB/s) to every Finnish household by 2015. It also discussed that the Australian government intends to spend the largest international expenditure of A$43 billion on improving the speed of Internet broadband connectivity; the US to spend US$9 billion; the UK to commit £850 million; Brazil US$7.3 billion; and other emerging economies such as India are also setting their national broadband plans due to the promise of the opportunities to be derived from connectivity. The Internet of Everything (IoE), therefore, is expected to impact on the economy by enhancing productivity, innovation and competitiveness through'automation and industrial manufacturing, logistics, business/process management and the intelligent transportation of people and goods' (Atzoria et al., 2011, p. 2787). Furthermore, it is expected to impact on the community through assisted living and e-health and e-education through synchronous engagement between teachers and students and will benefit the community through more vibrant living places and citizen participation (Atzori et al., 2011). The Internet played a role in the uprising in the Arab Middle East in 2011 (Dalacoura, 2012). Tufekci and Wilson (2012) outline how social media particularly through Facebook influenced the decision for individuals to participate the Egypt's protests. It is, therefore, associated with major changes. There is, however, no effective and significant framework with which to actually examine the impact of such connectivity on enterprise and community. However, the development of an impact framework is significant, given the broad agreement that measuring the impact of the Internet is vital for ensuring the success of such major digital infrastructure initiatives (DBCDE, 2011; European Commission, 2009; Firth and Mellor, 2005; US Broadband Plan, 2010).

The development of such a framework is significant for many reasons. Although the primary reason why such empirical models do not exist concerns the difficulty in defining impact and in sourcing sectoral-level data, there is increased recognition internationally about the importance of developing a framework to assess the economic and social impact of the Internet (European Commission, 2009; US Broadband Plan, 2010). The European Commission (2009) recognizes that 'there is a clear need of keeping track of developments in the use of ICT technologies that have recently attracted policy interest (p. 15). The EC also emphasizes that the role of ICT as a driver of process, product and business model innovation is largely unexplored at the empirical level' ...and also stresses that the 'social impact of ICT is still basically unexplored' (p. 16). They emphasize the importance of looking at the impact on the 'life of citizens, seen as learners, workers, consumers, patients, communicators and creators, in addition to the role of ICT in promoting social capital (p. 17). The US Broadband Plan (2010) also highlights the importance about knowing how 'broadband affects the very core of the economy: innovation, productivity and the way people live and work... and the need to derive new indicators to measure changes in industries and activities that do not yet exist' (p. 335). It argues that 'faster broadband speeds and better broadband quality improve incentives for entrepreneurs to innovate and savings realized by incorporating broadband into existing areas like education and health care' (p. 335). Australia is essentially part of a wider world Broadband Plan movement; however, like many nations such as the US, the promotion of such plans may be driven by perception rather than real understanding of Internet impacts.

The proposed research, therefore, focuses on developing a framework for assessing the impact of the IoE on key sectors including business, education, health and community.

THEORETICAL BACKGROUND

The Internet literature has largely ignored the critical focus on impact. The current literature predominantly emphasizes macro-level access factors measured by broadband coverage, penetration rates, connection types and user demographics, and high-level broadband policy development and instrumentation (Madden et al., 2000; Preston et al., 2007; Ramírez 2007; Tookey et al., 2006). For example, general broadband research using macro-economic country-level statistics has identified numerous factors which can be categorized as demographic (e.g., income, education, population density), marketing (pricing policies, e.g., bundling), economic (GDP, competition, government policy including subsidies), and infrastructural (e.g., national preparedness including underlying broadband infrastructure, Internet access, user skills) (Cava-Ferreruela et al., 2006; Distaso et al., 2006; Garcia-Murillo 2005).

There is research into impact, although scattered and largely non-systematic. At the macro-level, the literature has a predominant focus on GDP and other measures of economic growth (Lehr et al., 2006). Moving beyond the macro-level, there is strong recognition that sectoral-level analysis with the use of surveys is necessary to measure the impact of the Internet (EC 2009). There are also anecdotal studies that question the so-called beneficial impact of the Internet and call for more rigorous research to examine such impact. Firth & Mellor (2005) argue that the Internet literature is largely based on rhetoric and is rarely directed at the social impact of the Internet. They question the espoused benefits of the Internet as they raise controversies in terms of social relations as 'poorly socialized individuals may continue to be excluded; institutions may increasingly replace personal contact; and decision making by authority may increasingly be influenced by additional participation from a decreasing minority of citizens' (p. 11). Additionally, they also examine the health perspective and raise questions about the misinformation and inappropriate diagnosis involved in e-health. In terms of education, they caution against the temptation to standardise education with stifled choice and creativity. Therefore, more research is needed to move beyond the macro-level to the deeper sectoral levels.

This study aims to address these shortcomings. It advances knowledge by contributing a framework to measure the impact of the Internet on business and community. Current studies generally focus on technology adoption from an individual perspective through theories such as the Technology Acceptance Model (TAM) (Davis, 1989) and the Unified Theory of Acceptance and Use of Technology (UTAUT) (Venkatesh et al., 2003); an organizational perspective through the technology-organization-environment (TOE) framework (Depietro et al., 1990) and a national level with a focus on macro factors (Cava-Ferreruela et al., 2006; Distaso et al., 2006; Garcia-Murillo 2005). However, more research is needed on the sectoral level (Chiasson et al., 2005). In particular, extending existing research concerning enhancing understanding of how technologies such as the Internet are adopted and impact on key sectors including business, education, healthcare and community development will be useful in ascertaining whether key sectoral broadband benefits can be achieved. More broadly, information systems scholars and theorists have argued vigorously concerning the need for taking sectoral differences 'seriously' in adoption research and limited availability or even lack of micro-level sectoral data has much to answer for the identified shortcomings of extant adoption research (Chiasson et al., 2005).

The literature generally focuses on macro-economic factors rather than business-level factors. There is widespread agreement by both national governments and international economic organizations such as OECD, EU and UN that broadband and associated infrastructure can result in accelerated national economic growth and increased productivity. Early evidence from leading broadband economies such as South Korea, Japan, Canada, Sweden, Denmark, Netherlands, Iceland, and Switzerland supports these claims (Lee, 2007). Additionally evidence from the US and Australia confirms these findings and adds employment growth and labour productivity growth (Lehr & Osorio, 2005) in a number of business sectors generally, and particularly in businesses that operate in IT-intensive sectors (Lehr & Osorio, 2005). In addition to visible broader economic impacts and productivity improvements, the impact of broadband adoption and diffusion has been observed in other areas leading to economic development and value such as knowledge creation and dissemination, creation and development of strategic alliances in the hospitality industry, and outsourcing of business services and process operations (OECD, 2003). An under-represented area is the impact on business innovation (European Commission, 2009). The current Internet benchmarks simply assess innovation by the level of R&D expenditure. Additional emphasis should also be placed on the impact on innovation and its tangible and intangible outcomes rather than merely R&D expenditure which is not necessarily a true and sole measure of innovation.

Extant literature also provides mixed results on the impact of the Internet on commercial real-estate, which is a factor that affects business costs. On one hand, new working practices such as outsourcing, hot-desking, delayering, and corporate downsizing can intensify downward pressure on commercial real estate (Sing et al., 2006). On the other hand, when compared to unconnected commercial real estate, offices that are broadband-ready are likely to attract a premium relative to those that are not (Sing et al., 2006; Wong, 2000). This premium can be driven by quicker leasing, higher rental cash flows which may include occupation charges for broadband services provided (Spurge & Almond, 2004). As part of new working practices, there is a growing trend in teleworking or working from virtual offices (Spurge & Almond, 2004). There is general agreement that the issue of ICT impact on commercial real estate demand and value both in metropolitan and regional areas remains unresolved and there are growing calls for further research in this area (Sing et al., 2006; Spurge & Almond, 2004). The extent to which the Internet influences education is a contentious issue. On one hand, there is the lure of the Internet as a panacea to improve the quality, availability and reach of education, regardless of economic, social and geographic proximity to schools. On the other hand, questions arise over the quality of teaching materials, their appropriateness to an online environment and the extent of its usefulness to asynchronous education (downloading of teaching materials) in comparison to synchronous education (real-time engagement which is more useful) (Firth and Mellor, 2005). Furthermore, there is cynicism against the value of the Internet is substituting other important delivery modes which are critical in the transfer of tacit knowledge (Kiernan 2002) and experiential learning essential in developing innovation skills which are needed in the 21st century (Green et al., 2002). Similar to education, the literature contains contradictions on the impact of the Internet on health. The Internet claims that it would assist providers in the healthcare sector in diagnosing and monitoring diseases and offering remote care to patients, particularly in regional communities by reducing time and costs associated with travel and essentially saving lives given access to specialised services. There is evidence that the Internet may have an impact on tele-radiology given the high level of bandwidth required (Wright 1992). However, there is little evidence on the impact of the Internet on tele-medicine given issues associated with policy, privacy and practices (Whitten et al., 1995). The Internet has the potential to influence community development through social inclusion with the aged, disabled and socially excluded. Although 'community' is a dynamic concept there is agreement that it can be defined both in terms of locality (Quan-Haase et al., 2002), and also as 'online networks' or 'virtual communities' or 'social networks' that "provide sociability, support, information, a sense of belonging and social identity" (Quan-Haase et al., 2002, p. 2). The Internet can have an impact on community development. For instance in South Korea, the Internet has played an important role in fostering improved community engagement among excluded groups such as the elderly, housewives and farmers (Lee et al., 2003). More extensive research is needed on the impact of the Internet on the community through improved access to education, health and community services to allow policy makers to best leverage the capacity of the Internet as a positive force in society.

METHOD

This exploratory study is based on a case study, an empirical inquiry that attempts to investigate a problem in its real-life context. A case study approach is appropriate for achieving the goal of this study as it captures "'reality' in considerably greater detail (and the analysis of a considerably greater number of variables)" (Galliers 1990, p. 162). This case study was conducted in Willunga, a regional community in Australia where high-speed broadband Internet was rolled out in 2011. It was the first release site in the State for the National Broadband Network (NBN), an initiative by the Australian government to make fast Internet available to the entire country.

In-depth face-to-face interviews were conducted with key informants and were seen as valuable as information was expected to vary considerably (Ticehurst et al., 2000). Moreover, interviews are flexible, in that they empower interviewers to control the sequencing and wording of questions, although also providing opportunities for seeking clarifications (Walsham 1995). Interviews focused on the current and anticipated impact of the Internet and how to best assess it. Interviewees were first identified via snowballing whereby the leading government officials in the Economic Development Board for the region and subsequent interviewees referred the researchers to key informants in the region including leaders of business associations; small and micro businesses involved in food and wine, manufacturing, medical equipment, real-estate; local government Councils; and e-education specialists. Table 1 provides the details for each interviewee. The data collected from the interviews were analysed thematically. Codes were developed as patterns emerged which assisted in identifying and analysing unfolding themes (Carson et al., 2001). Data relevant to each theme were subsequently assembled before triangulation against secondary data and extant literature was carried out and representative quotations short-listed for support. Construct validity has been adequately addressed. The chain of evidence, tracing conclusions to the collected data was also maintained (Yin 2003). This measure has enhanced research reliability, thereby improving its overall quality.

Table 1. Details of Interviewees

Interviewee #	Organization Type	Position	Areas
1	Government	General Manager	Exports
2	Government	General Manager	Small business development
3	Government	General Manager	Industry development
4	Government	Director	Economic Development
5	Government	Officer	Economic Development
6	Government	Officer	Economic Development
7	Government	Coordinator	e-Learning
8	Government	General Manager	ICT
9	Government	Coordinator	ICT
10	Government	Officer	Policy
11	Government	Business Partner	Business development
12	Business	Managing Director	Manufacturing
13	Business	CEO	Science and Engineering
14	Business	Proprietor	Wine / Hospitality
15	Business	CEO	Consulting and Training
16	Business	Managing Director	ICT
17	Business	Proprietor	Real-estate
18	Business	CEO	Consulting

Results and Analysis

This section validates the framework arising from qualitative data and literature.

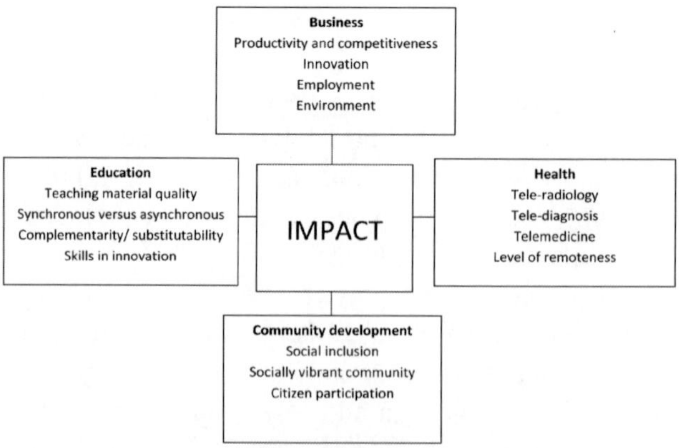

Figure 1. Framework for Assessing the Impact of IoE.

Figure 1 features the key issues from the framework which will be discussed further in this section.

Business

Interviewees (12 and 13) discussed the impact of the Internet on their firms' productivity and competitiveness. Interviwee12 indicated that his firm enjoyed export sales of over 60% of its sales from countries such as the UK which is its largest market followed by the USA and Germany and then Canada, New Zealand, South Africa and other European markets including Austria, Italy, France, Ukraine, Czech Republic, Norway and Finland. He indicated that the Internet was instrumental in reaching these markets in a cost effective manner.

He also described advantages of the Internet for innovation pertaining R&D and manufacturing firms. He explained how Internet technologies such as laser centering can be used for prototyping. It involves rotating the product virtually to develop a technical specification. Therefore, someone from overseas can design the product and developers can create the product physically based on those exact specifications. Thus, the Internet facilitates connection and communication with a wide international pool of expertise to enhance R&D efforts towards effective product development.

Interviewee 5 explained how broadband internet helped him and his firm through the innovation process as he uses the internet to research technologies and develop ideas for new products.

The interviewees also discussed the impact of broadband internet on employment and teleworking:

> And there are a lot of very creative, lateral thinking people, and a lot of entrepreneurial people in Willunga. So I think the [the Internet] will be really, really good for the town and many people, especially with their work. (Interviewee 17)

Key real-estate agents in Willunga pointed out that consumers are considering NBN access in the purchase of properties:

> Whether or not a property that they're looking at has got access to broadband: that's really important. And since the NBN is starting in Willunga, sellers of property are recognising the NBN as an asset and advertising that we've got the NBN... I think most purchasers would want it

just because of the speed. Yeah, I think somebody who owns rural properties just out of Willunga would be really, really disappointed that they are not going to be able to have it. (Interviewee 17)

Interviewee 7 also discussed the impact the broadband internet can have on the environment in reducing the amount of printed materials by providing access to materials electronically:

We encourage reading online rather than printing everything, so we try and keep printing to a minimum (Interviewee 7).

Community

Education

Qualitative data supported the necessity for both students and educators to have Internet access in their homes and schools respectively for synchronous education, provided that there are quality teaching materials and resources to support the process (Interviewee 13).

In cases of limited access, the delivery of education was problematic as Interviewee 15 expressed how students in Port Pirie could not access online materials due to poor access in regional Australia. Additionally, it should be carefully considered whether the online mode of delivery is complementary and not a substitute for other important modes of delivery for the transfer of tacit knowledge.

Health

The qualitative data confirmed the importance of the Internet on the health sector to some extent. On one hand, interviewees confirmed that the Internet may impact on tele-radiology: *We need lives saved by medical scanning technology in remote environments (Interviewee 11).* Interviewees also felt that the Internet was usefulness for tele-diagnosis, contrary to the level of caution within the existing literature as they expressed that medical practitioners were reading people's faces and examining injuries via the Internet, thereby benefiting remote communities (Interviewee 2). However, a number of barriers can also affect the uptake of telemedicine which can, in turn, reduce the impact of the Internet. Interviewee 6 confirmed that the impact of tele-medicine may be affected due to policy issues, such as Industrial Relations Awards and working hours. Additionally, the level of remoteness of a community can also arise as a key issue in assessing whether the Internet will have an impact on health services as this may vary from urban to remote communities. Interviewee 2 confirmed sentiments in the literature that the Internet would particularly benefit remote communities (Wooton, 1999).

Community Development

Qualitative data confirmed the impact of the Internet on community development through social inclusion. In particular, Interviewee 13 described how the Internet has helped in Aged care complexes. Although some argue that 'poorly socialized individuals may continue to be excluded' (Firth and Mellow, pp. 223), interviewees suggested that the Internet can provide an avenue to foster a socially vibrant community, by listing community events on websites and mobile applications that can be accessed by everyone in the community including the aged, young mothers, new migrants and socially excluded (Interviewees 11 and 18). Interviewees expressed that the Internet may have a positive impact on community development through improved community participation. *Social media, particularly for us, I suppose, will be vital for community consultation... So many of our projects, we need that consultation and I think that's the best way to do it (Interviewee 6).* Although many argue that online participation simply mirrors offline participation and decision-making authority continues to be skewed towards a minority of citizens (Firth and Mellor; Katz and Rice, 2002), interviewees felt that the Internet will increase stakeholder participation. Therefore, it is important that appropriate policies be put in place to ensure that groups of citizens based on demographics of age, education and income, be given equal opportunities to participate in political and community decisions.

CONCLUSION

This chapter provides a framework to assess the impact of the Internet of Everthings on business and community. The research is important as it pertains to impact unlike the existing literature that mainly focuses on adoption. The framework contributes pertinent insights to policy makers and managers, focusing on harnessing the opportunities provided in the Internet age from implementing high-speed broadband infrastructure and policies to implementing solutions in business to capitalise on increased connectivity. For government agencies, this framework can assist in designing interventions for leveraging the Internet for regional outcomes pertaining to education, health and community development. Businesses may also deem this research valuable in devising strategies in key sectors and facilitating cross-sector comparisons to identify growth opportunities for connectivity solutions in various sectors. Hence, this chapter can provide relevant insights to both governments and businesses for developing strategies for leveraging the increased capacity of the Internet to identify opportunities and pitfalls for realizing its anticipated benefits.

Although this chapter offers important insights, further work is necessary. It was based on a single case study which was appropriate given the exploratory nature of the study and understanding complex considerations. Therefore, further research should test the framework quantitatively in other contexts and can also be based on a longitudinal study to assess impact through time. Nevertheless, the study is vital in leading solutions to capitalise of the opportunities provided by the Internet of Everything.

REFERENCES

ABS (2010) 8166.0 - *Summary of IT Use and Innovation in Australian Business,* 2009-2010. Australian Bureau of Statistics.

Anderson, B. (2008). The social impact of broadband household internet access, *Information, Communication & Society,* 11(1), 5-24.

Atzori, L., Iera, A., and Morabito, G. (2011). The Internet of Things: A survey. Computer Networks. 54, 2787-2805.

Bauer, J., Gai, P., Kim, J., Muth, T., and Wildman, S. (2002). *Broadband: benefits and policy challenges.*

Carson, D., Gilmore, A., Gronhaug, K., and Perry, C. (2001). *Qualitative Research in Marketing* Sage, London.

Cava-Ferreruela, I., and Alabau-Muñoz, A. (2006). Broadband policy assessment: A cross-national empirical analysis, *Telecommunications Policy,* 30(8-9), 445-463.

Chiasson, M., and Davidson, E. (2005). Taking industry seriously in information systems research, *MIS Quarterly,* 29(4), 591-605.

CISCO (2013). www.cisco.com. Accessed December 11, 2013.

Cummings, S., Heeks, R., and Huysman, M. (2003). *Knowledge and Learning in Online Communities in Development: A Social Capital Perspective*, Paper presented at the Development Informatics Working Paper Series.

Dalacoura, K. (2012) The 2011 uprisings in the Arab Middle East: political change and geopolitical implications. *International Affairs*, 88(1), 63-79.

DBCDE (2011). #AU20 National Digital Economy Strategy. Department of Broadband, Communication and Digital Economy.

Depietro, R., Wiarda, E., and Fleischer, M. (1990). The context for change: Organization, technology, environment, in: *The process of technological innovation,* L.G. Tornatzky and M. Fleischer (eds.), Lexington Books, Lexington, MA, 151-175.

Distaso, W., Lupi, P., and Maneti, F.M. (2006). Platform competition and broadband uptake: theory and empirical evidence from the European Union., *Information Economics and Policy,* 18(1), 87-106.

EC (2009). *European Commission i2010 High Level Group: Benchmarking Digital Europe 2011-2015 a conceptual framework' European Commission: Information Society and Media.*

Firth, L., and Mellor, D. (2005). Broadband: benefits and problems, *Telecommunications Policy* (29), 223-236.

Galliers, R.D. (1990). Choosing appropriate information systems research approaches: a revised taxonomy, in: *The Information Systems Research Arena of the 90s. Perceptions and Alternative Approaches,* H.E. Nissen, H.K. Klein and R. Hirscheheim (eds.), Copenhagen.

Garcia-Murillo, M. (2005). International broadband deployment: the impact of unbundling, *Communications & Strategies,* 57(1), 83-108.

Goodhue, D.L., and Thompson, R.L. (1995). Task-technology fit and individual performance, *MIS Quarterly,* 19(2), 213-236.

Green, D., and O'Brien, T. (2002). The Internet's impact on teacher practice and classroom culture, *T.H.E. Journal,* 30(11).

Kiernan, V. (2002). Students embrace the Internet, but not as a replacement to conventional classrooms, *Chronicle of Higher Education,* (49(5), 45.

Kraut, R., Mukhopadhyay, T., Szczypula, J., Kiesler, S., and Scherlis, B. (1999). Information and communication: alternative uses of the Internet in households, *Information Systems Research,* 10(4), 287-303.

Lehr, W. H. & Osorio, C. A. (2005). Measuring broadband's economic impact. Available at: http://www.broadbandproperties.com/2005issues/ dec05 issues/Measuring%20Broadband%20Eco%20Impact,%20Lehr,%20Gilett, %20Sirbu.pdf. Broadband Properties Magazine(December), 12-24.

Lee, H., O'Keefe, R.M., and Yun, K. (2003). The growth of broadband and electronic commerce in South Korea: contributing factors, *The Information Society,* 19(1), 81-93.

Lee, S. (2007). The Determinants of the Global Broadband Deployment: An Empirical Analysis. Available: http://www.ptc.org/ptc08/participants/ speakers/papers/SangwonLee_FullPaper.pdf, Accessed on 1 June 2011.

Paper presented at the Pacific Telecommunications Council's 08 Conference.

Madden, G., Savage, S.J., Coble-Neal, G., and Bloxham, P. (2000). Advanced communications policy and adoption in rural Western Australia, *Telecommunications Policy,* 24(4), 291-304.

OECD (2010). OECD Information Technology Outlook 2010.

Preston, P., Cawley, A., and Metykova, M. (2007). Broadband and rural areas in the EU: From technology to applications and use, *Telecommunications Policy,* 31(6-7), 389-400.

Quan-Haase, A., and Wellman, B. (2002). How does the internet affect social capital, in: *IT and Social Capital* M. Huysman and V. Wulf (eds.).

Ramírez, R. (2007). Appreciating the Contribution of Broadband ICT With Rural and Remote Communities: Stepping Stones Toward an Alternative Paradigm, *The Information Society,* 23(2), 85-94.

Rampersad, G., and Troshani, I. (2011a). Bifocal technology adoption: A case study of a technology transfer tool, Pacific Asia Conference on Information Systems (PACIS), Brisbane, Australia.

Rampersad, G.C., Plewa, C., and Troshani, I. (2011b). Investigating the use of information technology in managing innovation: A case study from a university technology transfer office, *Journal of Engineering and Technology Management,* 29(1), 3-21.

Sing, T. F., Ooi, J. T. L., Wong, A. L. & Lum, P. K. K. (2006). Network connectivity and office occupiers' space decision: the case of Suntec City. Journal of Property Investment & Finance, 24(3), 221-238.

Spurge, V. & Almond, N. (2004). Broadband technology: How developers are responding to office occupiers' needs. Property Management, 22(2), 108-126.

Stone, R.W., Good, D.J., and Baker-Eveleth, L. (2007). The impact of information technology on individual and firm marketing performance, *Behaviour & Information Technology* 26(6), 465-482.

Ticehurst, C.G., and Veal, T.R. (2000). *Business Research Methods: A Managerial Approach* Pearson Education Pty Limited, Australia.

Tookey, A., Whalley, J., and Howick, S. (2006). Broadband diffusion in remote and rural Scotland, *Telecommunications Policy,* 30(8-9), 481-495.

Troshani, I., Rampersad, G.C., and Plewa, C. (forthcoming). Adopting Innovation Management Software in University Innovation Commercialization, *Journal of Computer Information Systems.*

Tufekci, Z. and Wilson, C. (2012) Social media and the decision to participate in political protest: Observations from Tahrir Square. *Journal of Communication.* 62(2), 363-379.

US Connecting America: *The National Broadband Plan.*

Venkatesh, V., Morris, M.G., Davis, G.B., and Davis, F.D. (2003). User acceptance of information technology: Toward a unified view, *MIS Quarterly,* 27(3), 425-478.

Walsham, G. (1995). Interpretive case studies in IS research: nature and method, *European Journal of Information Systems* (4), 74-81.

WEF World Economic Forum Global Information Technology Report 2010 - 2011: Transformations 2.0.

Whitten, P., and Allen, A. (1995). Analysis of telemedicine from an organisational perspective, *Telemedicine Journal,* 1(13), 203-213.

Wright, D.J. (1992). Strategic impact of broadband telecommunications in insurance, publishing, and health care, *IEEE Journal on Selected Areas in Communications* 10(9), 1369-1381.

Yin, R.K. (2003). *Case Study Research: Design and Methods*, (Third Edition ed.) Sage Publications, Thousand Oaks, London, New Delhi.

In: Technology Innovation Leadership … ISBN: 978-1-63117-388-2
Editors: G. Rampersad and F. Patel

Chapter 8

EVALUATING THE SUCCESS OF ICT IN SMEs IN DEVELOPING COUNTRIES

Amal Alshardan**, *Robert Goodwin* and *Giselle Rampersad
Flinders University, Australia

ABSTRACT

In recent years, there is a growing propensity in small and medium enterprises (SMEs) in developing countries to invest in Information and Communication Technology (ICT) expecting positive gains to their organizations. Although much research has been conducted on the benefits of ICT on both large and small organizations mainly in developed countries, there is a dearth of research on the benefits of ICT for SMEs in developing countries. SMEs in developing countries purportedly have substantially different characteristics, needs and resources compared to organizations of all size in developed countries. This research aims to develop a preliminary benefits measurement model for ICT in SMEs in developing countries. The preliminary framework, which reflects the characteristics of SMEs and developing countries, is derived using content analysis of 20 case studies from developing contexts including Lebanon and Turkey. Three hundred and forty identified benefits were then synthesized into sixty-four non-overlapping dimensions connected to five key areas of individual impact,

* Corresponding author: alsh0249@flinders.edu.au.

organizational impact, system quality, information quality and vendor quality. This research contributes a vital IS-Impact measurement model for use in developing countries.

INTRODUCTION

In the increasingly competitive global market, small and medium enterprises have frequently been the engine of economic growth and prosperity (Vrgovic, Glassman, Walton, & Vidicki, 2012). Small and medium enterprises (SMEs) represent the majority of companies operating in the world in both developed and developing countries. The use of information and communication technology (ICT) in enterprises is paramount to international competitiveness. ICT can bring benefits in terms of efficiency, effectiveness and innovation (Plewa et al., 2012; Plewa et al., 2011; Rampersad et al., 2012). However, there are many challenges facing SMEs when it comes to successfully implementing and using ICT mainly due to resource and skill limitations.

Although the social and economic benefits of ICT are well publicized, the usage of ICT among SMEs in developing countries is still considered embryonic (Patel, Sooknanan, Rampersad, & Mundkur, 2012; Wei, Loong, Leong, & Ooi, 2009). Moreover, studies in developing countries pertaining to IS evaluation is still immature and further understanding in developing countries by the wider information systems (IS) research community is necessary in order to proceed from basic to more elaborate and in-depth research accounts of ICT phenomena in developing countries (Avgerou, 2008). For good business performance, measuring and evaluating the success of ICT is critical for ensuring the on-going success of these investments and to avoid the risk of failure (Consoli, 2012).

In this chapter, we analyse the literature on success measures of ICT in SMEs in developing countries. We first review literature pertaining to SMEs, ICT and developing countries to understand the context of the research. We then review existing models used to measure the success of ICT. Content analysis is then applied to analyse 20 case studies of ICT for SMEs in developing countries, from which we derive a list of success measures. A further analysis of the measures with the characteristic of both SMEs and developing countries is presented and the result is combined with the previous studies and models to form the comprehensive model to measure the success of ICT in SMEs in developing countries.

SMALL AND MEDIUM-SIZED ENTERPRISES (SMEs)

SMEs exert a strong influence on the economies of all countries. They play an important role as a job creator and an innovative engine in the economy (Lin, 1998; Snider, da Silveira, & Balakrishnan, 2009).

As widely acknowledged by scholars, there is little consensus on a definition of SMEs (Gooding & Iii, 1985; Kimberly, 1976; Lee, Kim, & Kim, 2007). Existing definitions range from those based on quantitative measures such as number of employees, turnover or assets (Blau, 1970; Fathian, Akhavan, & Hoorali, 2008) to others which employ qualitative approaches reflecting development stages and organization strategies (Lee, et al., 2007).

CHARACTERISTICS OF SMEs

Despite the lack of universality of definition, it is important to understand that SMEs have special characteristics that differentiate them from large organizations. These characteristics play an important role in the adoption, use and management of ICT. Understanding SME issues and characteristics are crucial before making any attempt to measure ICT success (Petter, DeLone, & McLean, 2008). The unique characteristics of SMEs appear in many aspects within an organization. They are apparent in organizational structure, management and decision making. The characteristics also emerge in risk taking and product and service procedures (Blili & Raymond, 1993; Kartiwi & MacGregor, 2007; MacGregor & Vrazalic, 2006).

The organizational structure of SMEs is different from large organizations. SME structure is often a centralized informal structure (Ein-Dor & Segev, 1978; Kartiwi & MacGregor, 2007; Seibert, 2004; Snider, et al., 2009). Flexibility is perceived to be a positive characteristic of SME structure given the behavioural advantages derived from their greater flexibility and ability to adapt to changes in the market (Fathian, et al., 2008; Mc Cartan-Quinn & Carson, 2003). On the other hand, SME structure is criticised for being fragile (Snider, et al., 2009). This fragile structure of SMEs can be attributed to the poor management skills (Kartiwi & MacGregor, 2007). Furthermore, Guy and Joseph (1993) identified the difficulty of function segregation as an important differentiating characteristic of SME structure.

In extant SME research there is an agreement on the important characteristic of the lack of human and financial resources (Kartiwi & MacGregor, 2007 ; Mc Cartan-Quinn & Carson, 2003 ; Yap, Soh, & Raman, 1992). This characteristic directly affects many aspects of IS, such as training and consultancy support. The lack of financial resources is also related to the difficulty in obtaining credit (Snider, et al., 2009). SMEs are also more reluctant to spend on information technology and therefore have limited use of technology (Kartiwi & MacGregor, 2007). Levy and Powell (2000) found that SMEs may not be able to respond to the introduction of ICT due to their limited resources, including implementation and training. In addition, financial issues may lead to project delays or even abandonment (Bohórquez & Esteves, 2008 ; Snider, et al., 2009). There are many characteristics related to management, decision making and risk. SMEs have lower control over their external environment than larger businesses, and therefore are more vulnerable to uncertainty (Kartiwi & MacGregor, 2007). SME lack of knowledge and experience often means that they are faced with a high level of uncertainty in terms of the new technological environment and the use of information technology for strategic or competitive purposes (Blili & Raymond, 1993).

Owner-management is a common feature of SMEs (Yap, et al., 1992). Blili and Raymond (1993) discussed the dominant role of the owner which is also associated with limited information sharing and limited delegation of decision-making. Accordingly, SMEs seldom have senior management involvement in IS decisions (Levy & Powell, 2000; Snider, et al., 2009). Kartiwi and MacGregor (2007) indicate that SME owners often withhold information from colleagues. SMEs face more risks than large businesses resulting in high failure rates (Kartiwi & MacGregor, 2007). Hence SMEs are more reluctant to take risks or invest in information technology and therefore have limited use of technology. This feature has been proved by many researchers who have demonstrated that most SMEs avoided sophisticated software and applications (e.g., Chen, 1993; Cragg & King, 1993; DelVecchio, 1994; Holzinger & Hotch, 1993; Khan & Khan, 1992), which makes the propensity for SMEs to invest in IS much lower (Levy & Powell, 2000).

Lack of IT staff and departments is an important feature in SMEs (Guy & Harold Joseph, 1993). SMEs seldom have specifically designated IT employees or an independent formal IT department, instead depending more on external programming services and external expertise (Cragg & Zinatelli, 1995 ; Yap, et al., 1992; Levy & Powell, 2000; Snider, et al., 2009).

These characteristics make it clear that such an IS model is imperative to SMEs because they may become victims of fast and intuitive decisions.

Moreover they have a very high failure rate, so the risk on keeping unbeneficial ICT is very high.

INFORMATION AND COMMUNICATION TECHNOLOGIES (ICT)

As the name suggests, ICT involves technology that enables the processing, transfer and exchange of information and communication services in a digital form. It includes personal computers, digital television, email, mobile phones and the internet (Osterwalder, 2002).

The current business environment is becoming more competitive. With an expectation of better quality products in a shorter period of time, many companies increasingly are facing the challenges of operating efficiently. New information technologies can function as drivers to increase business efficiency and productivity in the current global economy (Grama & Fotache, 2007; Wei, et al., 2009). Moreover, detailed analysis of academic studies and experience around the world reveal an intense impact of ICT on improving the competitiveness of SMEs and achieving specific social and economic development goals as well as play a key role in broader national development strategies (Fathian, et al., 2008; Osterwalder, 2002; Rampersad, et al., 2012).

More precisely, if ICT is used appropriately, it will improve the efficiency and effectiveness of organizations by increasing value, automation, integration of business processes, sharing of data and practices, and real-time information (Wei, et al., 2009). In addition, ICT is fundamentally changing the nature of businesses as they open new productive perspectives (information integration), opportunities for innovation (the introduction of informational services), as well as new organizational and business opportunities (electronic trade, e-business platforms, inter-organizational networks) (Grama & Fotache, 2007).

ICT IN SMES

Despite the strong trend towards academic research in ICT related aspects in organizations, there is still a paucity of research on small to medium-sized enterprises (SMEs) and consequently, there is a low diffusion of ICT in SMEs. SMEs are not fully exploiting the potential of ICT to the extent of large companies. This is partly due to the fact that SMEs have limited resources, technology and capabilities (Consoli, 2012). It is also important to note that the benefits of ICT occur only after a period of adoption. To facilitate effective ICT investment by SMEs, it is important to evaluate these investments continuously to ensure success (Consoli, 2012).

ICT in Developing Countries

ICT in developing countries is still impractical to various groups due to a number of reasons. One of them is a lack of appropriate ICT products as existing products are often designed to meet the needs of developed countries (Berisha-Namani, 2009). Additionally, limited financial resources is another feature that exists in both SMEs and developing countries (Berisha-Namani, 2009). Furthermore, the lack of expertise and human resources in SMEs in developing countries is also problematic (Alghamdi et al., 2011; Berisha-Namani, 2009). Another challenge is the lack of robust regulatory frameworks for ICT in developing countries (Alghamdi et al., 2011).

Language barrier is also a consideration for developing countries especially when the appropriate ICT product may exist only in English-language which is a common problem in many ICT products. In developing countries, citizens may not necessarily know other languages beyond the local language (Grazzi & Vergara, 2012). As a result, studies reflect high failure rates of IS projects in developing countries due to language barriers (Heeks, 2002).

Given the vital role of ICT for economic growth in developing countries coupled with the high failure risk in these countries, it is important to examine the factors that contribute to the success of ICT in developing countries.

ICT in SMEs in Developing Countries

With the increasing importance of ICT and SMEs, many researchers have published empirical studies on different aspects of ICT in SMEs internationally. Most studies focus on the adoption of ICT in SMEs in different developing countries such as in Malaysia (Alam & Noor, 2009), Nigeria (Irefin, Abdul-Azeez, & Tijani, 2012) and in the Kingdom Of Saudi Arabia (Skoko, 2012; Skoko & Ceric, 2010).

Other researches have focused on the use and impact of ICT in SMEs in developing countries. For example, a study in Oman presents the results of an exploratory study carried out in 51 SMEs in Oman to learn about the use and impact of ICT. The study provides an insight on the barriers for the adoption of ICT including the lack of internal capabilities, high cost of ICT and lack of information about suitable ICT solutions and implementation (Ashrafi & Murtaza, 2008).

This result is consistent with a study in Kosova which also found similar factors that attribute to a low level of information technology usage in Kosova (Berisha-Namani, 2009). Another study examines the impact of ICT on SMEs based on the context of Qatar. The study identified the need for more training facilities and government support for adopting ICT in Qatar's SMEs (Manochehri, Al-Esmail, & Ashrafi, 2012). In Nigeria, another study argued that the Nigerian government must invest largely in the SME sector by making more funds available to SMEs and putting structures in place to ensure successful investment (Apulu & Latham, 2010). From a South African point of view, Modimogale and Kroeze (2011) explore the barriers of ICT usage in SMEs in Gauteng and list recommendations for SMEs to overcome barriers and improve implementation of ICT in their businesses such as educating their staff and management about ICT, investing in recruitment or outsourcing knowledgeable ICT specialists, and building a culture that is innovative and favourable to technology (Modimogale & Kroeze, 2011).

Few studies attempt to measure benefits of ICT in SMEs in developing countries (Ndiege, Wayi, & Herselman, 2012). Therefore, developing a measurement success model is essential for evaluating ICT in SMEs in developing countries in order to justify their value and contribution to productivity. Kate et al. (2010) surveyed 130 Indian SMEs to find out whether and how Indian SMEs are benefiting from enterprise resource planning ICT implementation.

That study revealed that most SMEs implemented ICT to integrate the existing information system and found ICT implementation mainly beneficial in reducing inventory, improving customer services and improving communications. Also the study found that top management support, and user involvement and participation are the major contributors of ICT success (Kale, et al., 2010). In Jordan, Hawari and Heeks (2010) developed a "design-reality gap" model and applied it to a case study of ICT failure in a Jordanian manufacturing firm. Analysing the situation both before and during ICT implementation through a mix of interviews, observation and document analysis, their research finds sizeable gaps between the assumptions and requirements built into ICT system design, and the actual realities of the client organization. Their model derived from different IS success measurement models which includes seven dimensions summarised by the ITPOSMO acronym (Information, Technology, Processes, Objectives and values, Staffing and skills, Management system and structure and Other resources (Hawari & Heeks, 2010).

Moreover, Ndiege, Wayi, and Herselman's (2012) study focuses on assessing the quality of the IS used by SMEs in Kenya (Ndiege, et al., 2012). They found low usage of IS within SMEs attributed to low level of IS skills by both the SMEs management and the IS users, and the poorly designed information systems that do not adequately address the needs of the SMEs. They also evaluated the quality of IS in SMEs in developing countries by applying D&M model (explained in the next section) and found them of poor quality (Ndiege, et al., 2012).

Similarly, in Malaysia, Wei et al. (2009) present a re-specification to the D&M model. They proposed a conceptual model resultant from a comprehensive review of IS success literature. Their result provides an expanded understanding of factors that measure ICT success and suggest ways to improve ICT usage (Wei, et al., 2009).

MEASURING THE SUCCESS OF ICT

Organizations of different sizes around the world are investing heavily in ICT (Petter, et al., 2008). Naturally, they are expecting positive impacts for their organizations. It is thus important for them to measure and examine the success and impact of such a significant investment.

Yet, in the academic literature, there is little exploration of the impacts of ICT. Gable et al. (2008) suggested five possible reasons for this issue: (1) incomplete or inappropriate measures of success, (2) lack of theoretical grounding of causal and process models of IS success, (3) myopic focus on financial performance indicators, and (4) weaknesses in survey instruments employed or data collection approaches. Thus, as ICT success measurement is very crucial to organizations, there is a strong need to develop an ICT success measurement model for different contexts to facilitate practice as well as research (Sedera & Gable, 2004). The following sections review important IS success models.

THE DELONE AND MCLEAN (1992, 2003)

The DeLone and McLean (1992, 2003) models are probably the most cited models in the IS community. DeLone & McLean (1992) carried out a review of research published during the period 1981–1987. Based on this review they created taxonomy of IS success. Thus, a full set of 119 success measures were summarised into six categories or components of IS success: system quality, information quality, use, user satisfaction, individual impact, and organizational impact. Figure 1 shows this original IS success model (DeLone & McLean, 1992).

This model suggests that when an IS system is first created, its features can be observed as some degree of systems quality and information quality, then users use the systems and are either satisfied or not satisfied with the systems, and the use of the systems will impact the individual's work performance, and consequently impact the organization (either positively or negatively) (DeLone & McLean, 1992).

Their 1992 model has been widely tested, modified and improved. Petter et al. (2008) reviewed 180 papers found in the academic literature for the period 1992-2007 dealing with some aspect of IS success in general and the DeLone and McLean models in particular. Pitt, Watson and Kavan (1995) evaluated the instrument from an IS perspective and suggested that the construct of service quality be added to the D&M model. Seddon and Kiew (1996) studied a portion of the IS success model and modified the construct, use, to usefulness.

Another proposed adjustment to the DeLone and McLean model is offered by Seddon (1997). He argued that the D&M model in its original form was confusing, partly because both process and variance models were combined within the same framework.

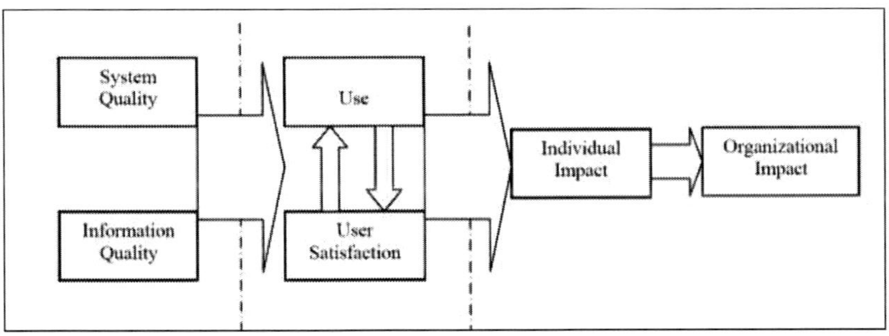

Figure 1. DeLone & McLean 1992 IS-Success Model.

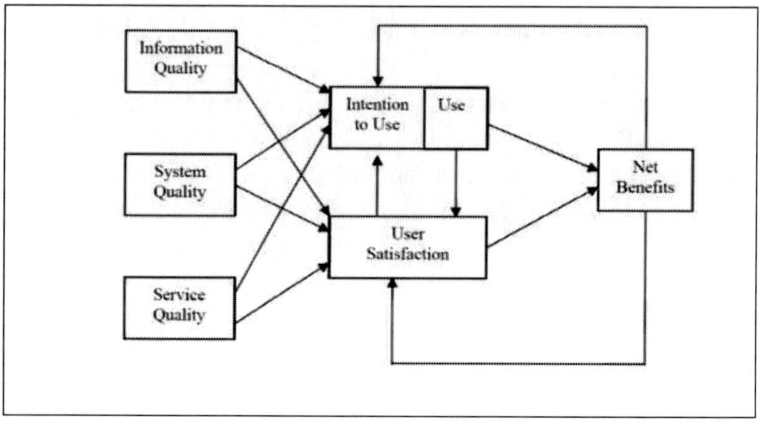

Figure 2. DeLone & McLean's 2003 IS-Success Model.

Seddon suggested that the concept of use is highly ambiguous and suggested that further clarification was needed for this construct. He derived three different potential meanings for the use construct, as well as parsing out the process and variances portions of the model. According to Petter et al. (2008), this change introduced by Seddon complicates the model, thereby reducing its impact.

Many researchers suggested revisions or extensions to the model in other contexts. Some have modified it to measure success of specific applications such as knowledge management (e.g., Jennex & Olfman, 2004; Kulkarni, Ravindran, & Freeze, 2007; Wu & Wang, 2006) and e-commerce (e.g., Delone & Mclean, 2004; Molla & Licker, 2001; Zhu & Kraemer, 2005).

Based on these improvements and other alternative frameworks for measuring IS effective, DeLone and McLean conducted an in-depth analysis, and they then updated their model and proposed a new DeLone and McLean's (D&M) 2003 IS Success Model (See Figure 2 above). There are three main enhancements in the updated IS Success Model 2003. Firstly, the updated IS success model accepted Pitt et al. (1995) recommendation to include service quality as a construct. Service Quality became a dimension of IS success not only a subset of "system quality" as before. The new model also addressed the criticism that an information system can affect levels other than individual and organizational levels including workgroups, industries, and even societies (Myers, Kappelman, & Prybutok, 1998; Seddon, Staples, Patnayakuni, & Bowtell, 1999). Consequently, DeLone & McLean replaced the variables, individual impact and organizational impact, with net benefits in that way accounting for benefits at multiple levels of analysis (Petter, et al., 2008). A third enhancement made to the updated DeLone and McLean model was a further clarification of the use construct (Petter, et al., 2008). The authors explained the construct as follows: "Use" must precede ''user satisfaction'' in a process sense, but positive experience with ''use'' will lead to greater ''user satisfaction'' in a causal sense' (DeLone & McLean, 2003).

The D&M model opened the gate for many researchers to either empirically test the model in different contexts or to critique the model or enhance it in some aspects. One of the criticisms of the D&M model is the lack of theoretical background. On this basis, Sabherwal, Jeyaraj and Chowa (2006) performed comprehensive meta-analysis examination to the D&M model.

According to Petter, et al. (2008), Sabherwal et al. (2006) work has been instrumental in synthesizing the quantitative research related to IS success and has validated a substantial portion of the D&M model. Sabherwal et al. (2006) provide insights into the success of an IS that is adopted or used by individuals within the organization, and the determinants of IS success, by empirically integrating prior research in this area. The study develops a comprehensive theoretical model, including constructs related to the context, the users, and IS success. The theoretical model they provided was tested through a combination of meta-analysis and structural equation modelling. The results underline the importance of user-related and contextual attributes in IS success.

The D&M 2003 model reflected the importance of "use" as a dimension of measuring IS success. However the authors mentioned the ambiguity of improper or simple definition of system use. Burton-Jones and Straub (2006) respond to D&M 2003 call and examined the system usage construct in more detail.

Their systematic approach for reconceptualising the system usage construct assists researchers in developing clear and valid measures of system usage for a given theoretical and substantive context. In their paper rather than onerously identifying an exacting and singular conceptualization of system usage, they present an approach for systematically developing conceptualizations of usage for specific contexts and selecting usage measures in a theoretically precise way. They defined system usage as "an activity that involves three elements: (1) a user, i.e., the subject using the IS, (2) a system, i.e., the object being used, and (3) a task, i.e., the function being performed.). Their approach consists of two stages: Defining System Usage and Selecting Content Valid. The later consist of a two-step approach including structure and function. (See Figure 3).

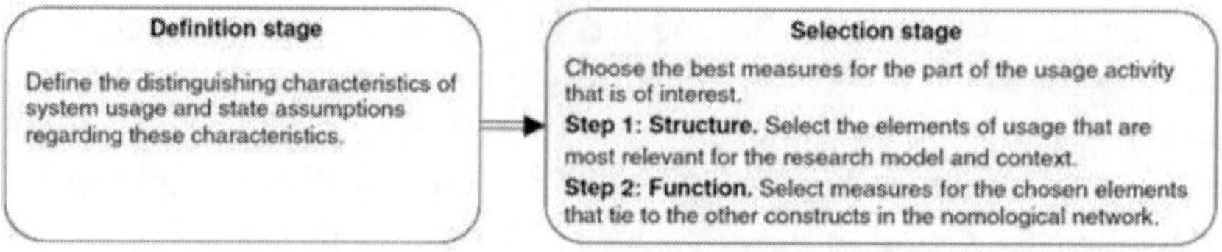

Figure 3. Staged approach for defining system usage and selecting usage measures (Burton-Jones & Straub, 2006).

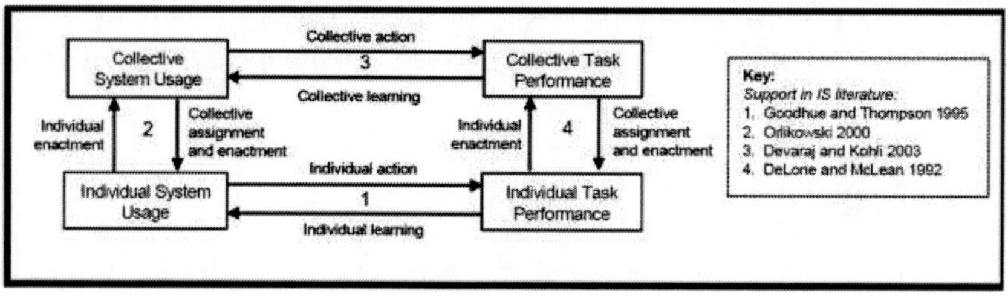

Figure 4. Theoretical Model of system usage (Burton-Jones & Gallivan, 2007).

In another study, Burton-Jones and Gallivan (2007) provide a multilevel theory of system usage. The study focuses on system usage multilevel nature. The authors believe that studying it one level at a time ultimately leads to an unnatural, incomplete, and disjointed view of how organizations function. Accordingly, they provide detailed steps for building multilevel theories of system usage, devised guidance for supporting each step, and provided a concrete illustration (see Figure 4)

THE IS-IMPACT MODEL

On the basis of DeLone & McLean's 1992 IS success model, and tests of other researcher's work (Myers, et al., 1998; Shang & Seddon, 2002), Gable et al. (2008) developed a more advanced IS-Impact model (See Figure 5). To obtain and validate this model, three surveys (referred to as identification, specification, and confirmation) were employed, with data gathered from a total of 600 respondents. Identification survey aims to identify the salient success dimensions and measures; Specification survey is used to specify the a-priori model using construct and measures identified; the purpose of confirmatory survey is to validate the a-priori model and instrument (Gable, et al., 2008). The method they used is a multi-method research design, extending the research cycle proposed by MacKenzie and House (1979) and McGrath (1979) for developing and validating a measurement model. It entailed two main phases and three surveys as shown in figure 5: (1) an exploratory phase, to develop the hypothesized measurement model, and (2) a confirmatory phase, to test the hypothesized measurement model against new data gathered.

The authors define the IS-Impact of an Information System (IS) as "a measure at a point in time of the stream of net benefits from the IS, to date and anticipated, as perceived by all key user groups" (Gable, et al., 2008, p.10). The IS-Impact Model is a holistic index representing the stream of net benefits; the impact half measuring net benefits to date, the quality half being our best proxy measure of probable future impacts, and "impacts" being the common denominator (Gable, et al., 2008). Three important issues are addressed in their paper to re-conceptualize the information system success as a formative construct; namely, completeness, mutual exclusivity, and necessity of dimensions and measures.

The IS-Impact model focuses on one casual flow of the IS net benefits, associating four dimensions with IT function. It consists of four constructs - Individual Impact, Organization Impact, System Quality, and Information Quality, representing four distinct but related dimensions of the multidimensional phenomenon: enterprise systems success, divided into two halves.

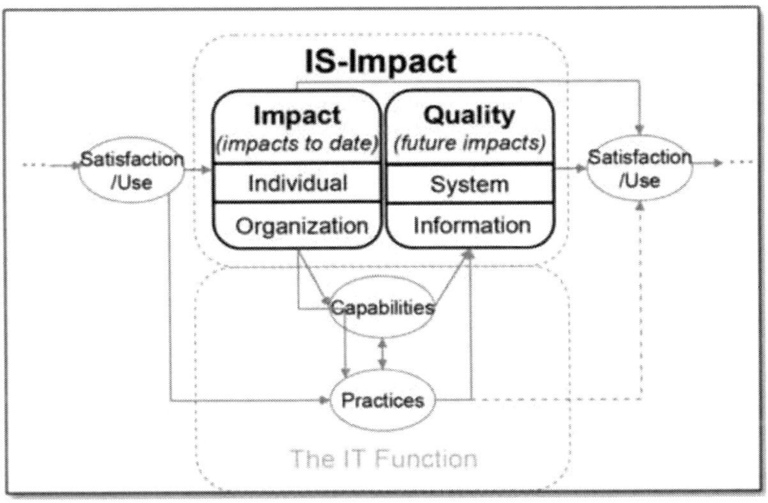

Figure 5. The IS-Impact Measurement Model.

Furthermore, the constructs of Use and Satisfaction have always been criticised by researchers for not being true measures of IS success. Through comprehensive exploratory studies, Gable et al. (2008) drew the conclusion: both Use and Satisfaction are antecedents or consequences of IS-Impact rather than two dimensions. When evaluating a system, measure of these dimensions represent results that are highly comparable across time, stakeholders, different systems, and systems context. The impact dimensions are an assessment of benefits of the system. The quality dimensions reflect future potential. Together, these four dimensions reflect a relatively complete view of the information system main measures of success (Gable, et al., 2008), see Figure 5.

Saunders and Jones' Model (1992)

Saunders and Jones (1992) developed the "IS Function Performance Evaluation Model" which was used to describe how measures should be selected from the multiple dimensions of the IS function relative to specific organizational factors and based on the perspective of the evaluator. Saunders and Jones conducted a Delphi study that examined how IS function performance dimensions were ranked in importance by IS executives, how the IS executive measures performed in each dimension, and the value they place on the measures. Through several interviews with chief executive officers (CEOs) and chief information officers (CIOs), they found the highest-ranked dimension was the IS function impact on strategic direction, followed by integration of the IS function planning with corporation planning, the quality of information outputs, and the IS function's contribution to organizational financial performance.

3-D Model (1996)

Ballantine et al. (1996) evaluated DeLone & McLean's (1992) model on a number of dimensions, and a new model-3-D model, was proposed which fundamentally extends DeLone & McLean's work. In this model, the concept of IS success was separated into three fundamental dimensions: the technical Development level, the Deployment to the user, and the Delivery of business benefits. As a further extension of DeLone & McLean's work, Ballantine et al.

(1996) work has the following four advancements. Firstly, the 3-D model does support more complex contingencies. Secondly, some progress is made in the confusion between dependent and independent variables. The influencing factors are the closest equivalent to independent variable; the outcomes of each level are the closest equivalent to dependent variables. Thirdly, the 3-D model recognises the needs of these stakeholders at different level. Lastly, they advocated considering different methods and measures in order to evaluate success at the different levels.

MYERS'S MODEL (1997)

Figure 6 shows the comprehensive, IS assessment framework proposed by Myers, et al. (1998). This IS assessment framework added two dimensions- Service Quality and Workgroup Impact, compared to DeLone & McLean's model (1992).

CAMERON & WHETTEN'S (1983) SEVEN QUESTIONS ON ORGANIZATIONAL EFFECTIVENESS MEASUREMENT

Seddon et al. (1999) recommended that, before seeking to evaluate an IT investment, it is necessary to have clear answers to each of Cameron & Whetten's (1983) seven questions on organizational effectiveness measurement. These questions are shown in Table 1.

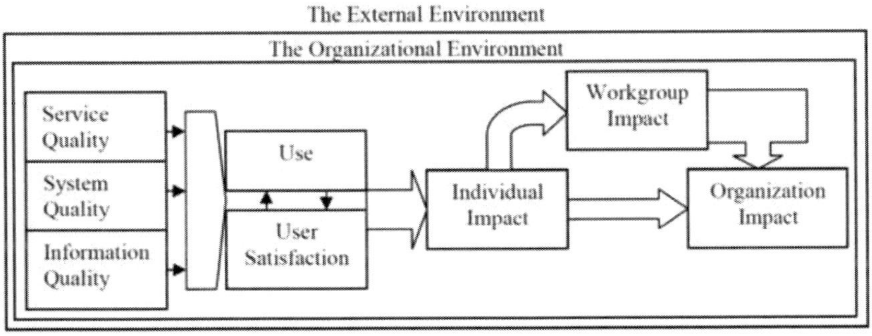

Figure 6. Myers's Model.

Table 1. Seven questions on organizational effectiveness -Cameron & Whetten's (1983)

seven questions on organizational effectiveness measurement From Cameron & Whetten's (1983)
1. From whose perspective is effectiveness being judged?
2. What is the domain of activity? (depends on tasks emphasized in the organization, competencies of the organization, and demands from external forces)
3. What is the level of analysis? (individual, subunit, organization, population, societal)
4. What is the purpose of evaluation?
5. What is time frame is employed? (short, long)
6. What types of data are to be used? (objective or perceptual)
7. Against which referent is effectiveness to be judged? (effectiveness of this organization)

DEVELOPING A MEASUREMENT SUCCESS MODEL FOR ICT IN SMES IN DEVELOPING COUNTRIES

Content analysis was used in this research to identify the benefits of ICT in SMEs in developing countries. The data source is customers' success stories in the commercial press of vendors' websites. The inclusion of commercial press as a source of evidence can be justified by several factors. First, ICT has recently been implemented in SMEs and its precise benefits have not been thoroughly examined or distinguished from large organizations by academic research. Second, the inclusion of benefits from quoted statement or video interviews represent the customer exact statement about the system. In addition vendors would have had to gain approval from customers to publish stories about them; it also anticipates that customers could be contacted directly to confirm details of the claimed benefits. Third, the first step in building the model is to include all the possible benefits, which is the main purpose for the vendors to gain the acceptance of customers for their product. Fourth, two leading' vendor's websites are examined to avoid the emphases on specific benefits related to specific products. Fifth, the development of a prior model will be guided by the previous studies and models and will be analysed and justified in the light of SMEs' characteristics and related developing countries characteristics.

THE CONTENT ANALYSIS PROCESS

We chose two leading companies in ES (ES represents the ICT for its integration and innovative IS technology), Microsoft for their solution for SMEs, Microsoft Dynamic NAV and SAP for three solutions for SMEs, namely, SAP business one, SAP business by Design and SAP business all in one.

The rational for choosing these vendors are based primarily on their leadership in ES in SMEs. Microsoft Dynamics NAV is a complete ERP solution for SMEs. It was established in 1984 and more than one million users have used Microsoft Dynamics NAV to simplify and streamline their business processes. Moreover, it is available in more than 42 different countries. SAP is a very large software company; it is been in business for more than 30 years. SAP has more than 97,000 total customers – 70,500 of which are small and midsize enterprises, and has customers in more than 50 countries. SAP has three different solutions for SMEs, SAP Business One, SAP Business ByDesign and SAP Business All-in-One. Depending on the SMEs needs and characteristics they choose the software adequate to their needs. We found that Microsoft Dynamic NAV and SAP customers' reports are more structured and contain traceable organizational details for future verification.

Using their websites, we employ the customers' "success stories" published on the Microsoft and SAP vendor website to develop a pool of benefits for ES for SMEs. The process of choosing the stories from these websites is random. The selected cases include different developing countries and different industries.

We identified keywords from the customer quotation and then derived the ICT benefits and linked it to a measurement dimension. As a result of these steps, a framework and pool of benefits of ICT in SMEs in developing countries is produced. Figure 7, illustrates the above steps and table 2 shows some examples of selected cases and the derived ICT benefits associated with the customers' quotations.

The content analysis of 20 published case studies containing 207 customers' quotations yielded 340 identified benefits which were then synthesized in to 64 non-overlapping benefits. Table 3 shows this pool of benefits. Processing the ICT benefits by removing duplication and combining similar benefits resulted in the following pool of benefits:

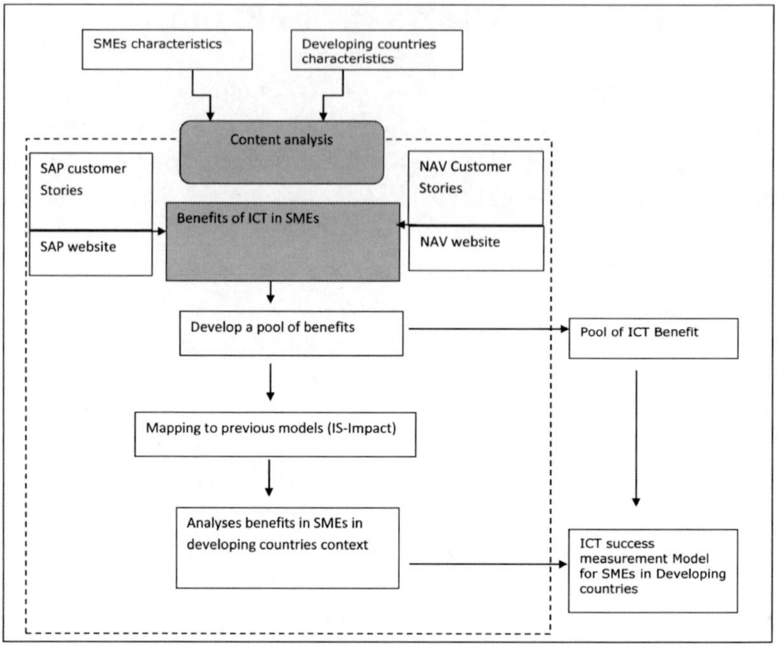

Figure 7. Methodology.

Table 2. Examples of ICT benefits

C#	country	Full quotation	ICT benefits
1	Turkey	"The solution is saving between 25 and 30 person days a month in payroll, accounting, and contract management."	employee effectiveness
2	Nigeria	"The rate of adoption of the software by employees was much faster than expected"	easy to learn
3	Lithuania	"Ultimately, we wanted a unified business management system that could support sales, finance, and human resource activities."	unified (integration)
4	Lebanon	"The application needed to be flexible and easy-to-use, so that we could modify it to suit an educational environment"	flexible - easy to use-customization
5	UAE	"Now, the company is well prepared to meet its global expansion plans"	support growth plans

Table 3. Pool of ICT benefits

ICT benefits				
improve individual productivity	better visibility	generate more revenue	compatibility	reliability
improve individual efficiency	Business Process Change	improve overall efficiency	Integration	sophistication
Individual learning	business process improved	improved overall productivity	Easy to learn	system accuracy
Individual self-confidence	centrally controlled	increase performance	data currency	multiple languages and currencies
accurate information	clear focus	increased capacity	easy to use	access to expertise
access to information	cost effective	improved outcomes /outputs	easily scalable	affordable
real -time information	cost reduction	organizational cost	familiarity	easy installation
better information	Improve administrative	Provides a Platform for the Future	flexibility	easy to support
critical information	enhance customer services	strengthened us	Improves security	Vendor locally available
Timeliness	enhance information flow	customization	meet company requirements	Vendor support
Importance	ensure sustained business growth	address all function	meet international standards	Vendor training support
valuable information	fewer employees	automated	powerful system	vendor popularity
decisions support	focus on key growth markets	better database Management	Quick data availability	
Total : 64 ICT Benefits				

MAPPING AND ANALYSING THE BENEFIT CITATIONS INTO PREVIOUS STUDIES AND MODELS

After the process of identifying the salient benefits of ICT in SMEs, the next step is to map those benefit statements into a conceptual foundation stemming from prevailing IS-impact theories. The main objectives of the mapping exercise are two-folded: (1) to provide a basis for the intended research, and (2) to demonstrate the possible inadequacies of existing models to measure the success of ICT in SMEs in developing countries.

The literature suggests two main approaches in developing an a-priori model: (1) a 'bottom-up', data driven, open coding approach or (2) a 'top-down', structured coding, framework approach (Gable et al., 2008). The top-down approach employs deduction, and starts with a logical framework or model to categorize the responses, whereas the bottom-up approach employs induction, starting with the data in hand, that is arranged into a logical classification. By using Gable et al. IS-Impact model as the theoretical foundation, this study employs a top-down approach. However, a bottom-up approach is also used for constructing the unmapped benefit citations.

In this research, the process of developing the success model for ICT in SMEs in developing countries follows the guidelines suggested by Gable et al. (2008): (1) model completeness - include all relevant dimensions and measures, where any ill-conceived additions or omissions good and bad, high and low, positive and negative may critically mask, neutralize or distort results, (2) model parsimony - where only the simplest and smallest relevant dimensions and measures are included, and (3) mutual exclusivity - where each measure addresses a unique benefit of ICT in SMEs without having overlapping measures. Moreover, Gregor (2006) proposed three points in the guideline of the Analytic Theory. Such guidelines are "…the logic for the placement of phenomena into categories should be clear, as should be characteristics that define each category. In addition, important categories or elements should not be omitted from the classification system, that is, it should be complete and exhaustive".

To make the mapping process clear, complete and exhaustive, we carefully introduced a mapping procedure to map the pool of benefits into the 4 dimensions and 37 measures of IS-Impact Model. The following is the suggested mapping procedure:

- Have these keywords and synonyms in the pool of benefits framework checked by a language specialist and IS specialist.
- Write synonyms for the IS-Impact keywords
- Classify all the measures in the IS-Impact and Benefit framework according to the level of measure into measure level and sub measure level
- Map them to the IS-Impact (same keyword or synonym) with the consideration of measure level, where some measures are set in a very high level whereas others are set in a more detailed level. For

instance, "improve business process" is a higher level measure and "improve accuracy of business process" can be considered a sub measure level.

- Double check on the unmatched measures by referring to the original quotations.
- Check the appropriation of adding them by checking the SMEs and developing countries characteristics.
- Classify the unmatched measures into groups
- Determine which dimension they belong to otherwise create a suitable dimension
- Add the unmatched benefit to the prior model
- Critically analyse the IS-Impact measures that have not been matched to any benefit from the pool against SMEs characteristics and developing countries characteristics. Determine whether they should be deleted or kept.

The mapping result is shown in Tables 4 and 5 below.

Table 4. The mapped IS-Impact benefits

Mapped IS-Impact benefits			
Individual-Impact	Organizational Impact	System Quality	Information Quality
Learning(1) Awareness - Recall(0) Decision-effectiveness(5) Individual-productivity(4)	Organizational-costs(1) Staff-requirements(2) Cost-reduction(17) Overall productivity(10) Improved outcome(3) Increased-capacity(5) e-government(0) Business Process change(2)	Data accuracy(1) Data currency(1) Ease of learning(9) Ease of use(11) Access(0) requirements(4) System feature(1) System accuracy(7) Flexibility(11) Reliability(5) Efficiency(1) Sophistication(2) Integration(36) customisation(16)	Importance(1) Availability(5) Usability(0) Understandable(0) Relevance(0) Content accuracy (5) conciseness(0) Timeliness(18) Uniqueness(0)
10	40	105	29

Table 5. The unmapped ICT benefits

Unmapped benefits				
Individual-Impact	Organizational Impact	System Quality	Information Quality	Vendor Quality
improve individual efficiency (6) Individual self-confidence (1)	better visibility (3) Cost-effective(4) Improve Business Process (2) centrally controlled(8) clear focus(1) Improve administrative(5) enhance customer services (2) enhance information flow(1) ensure sustained business growth(2) focus on key growth markets(5) generate more revenue(1) improve overall efficiency(24) strengthened the organization(1)	automated(14) better database Management(1) compatibility(1) Improves security (3) familiarity (3) easily scalable (15) Support multiple languages and currencies (1) Address all functions(1) meet international standards(6) powerful system(4) Quick data availability(4)	valuable(1) better(1) critical(1)	access to expertise(2) affordable(3) easy installation(20) easy to support(2) vendor locally available (1) Vendor support(5) Training support(3) vendor popularity(1)
7	59	53	3	34

The mapping procedure of the 340 measures to the IS-Impact shows 184 mapped measures and 156 unmapped measures. Unmapped measures from the IS-impact are critically analysed in view of SMEs' and developing countries' characteristics and accordingly were subject to be kept or removed from the model. As the purpose is to collect a comprehensive wealth of benefits, we were on the side of keeping the benefits unless it is completely unrelated to current SMEs and developing countries characteristics in which keeping them will just complicate the model. This is applied for example to E-government (organizational impact) and Learning and Awareness/Recall (individual impact).

A key observation herein is that the unmapped citations are largely related to the benefits associated with vendor/supplier. Given the citations are strongly emphasizing the linkages between organization and the support from the vendors/suppliers this dimension is tentatively labelled as "Vendor Quality".

Considering the D&M revised model (DeLone & McLean, 2003) the Service Quality is added as a new dimension of IS success as a replacement for being only a subset of "system quality". However this dimension does not exist in the IS-Impact model. Gable et al. (2008) justified,"... as the unit of analysis herein is the Information System (the ES), not the IT function, Service Quality was considered inappropriate" (Gable, et al., 2008, p. 13). However we found that many benefits in the context of SMEs in developing countries are related to this dimension, hence we added it as a fifth dimension in the proposed model.

We can infer form the mapping process that all the changes are related in some way to one or more characteristics of SMEs or developing countries. For example, the addition of scalability as a measure for system quality reflects the growth seeking characteristic of developing countries (Chinn & Fairlie, 2006; Vrgovic, et al., 2012). Similarly, the existence of support of multiple languages measure in system quality reflects the language barrier in adopting ICT in developing countries (Grazzi & Vergara, 2012). Another observation is the "centrally controlled" as an impact of ICT on the organization which directly reveals the organizational structure of SMEs, which are often centralized (Ein-Dor & Segev, 1978; Kartiwi & MacGregor, 2007; Seibert, 2004; Snider, et al., 2009). Also, it is clear that the technology and skills limitation of SMEs have a great impact on how they express or define benefits. For example "better information" is an expression that would include all the measures that relate to the information quality dimension such as importance, usability, content accuracy, conciseness, timeliness and uniqueness. Similarly, "strengthened the organization" is an expression for many positive impacts in the organizational level such as productivity, increased capacity, and improved outcomes.

Automation is a frequently reported benefit. This is mainly because the usage of ICT within SMEs was primarily for automation (Ndiege, et al., 2012), which reflects that most SMEs were reliant on manual tasks before adopting ICT. For the simplicity of the model, we also combined some different measures with minor differences representative of SMEs in developing countries. This includes productivity, efficiency and performance; data accuracy and content accuracy; data currency, timeliness; and clear focus, improved control, improved administration and some other similar measures in the system and information quality dimensions.

A final observation to address is the decision effectiveness measure. We move it from individual impact to organizational impact to reflect the centralized structure and the decision making characteristics of the SMEs

(Yap, et al., 1992). In SMEs there is a dominant role of the owner with limited information sharing and limited delegation of decision-making (Blili & Raymond, 1993). Also the "Business Process change" measure in the organizational impact stated by some cases as "Improve Business Process", because the Impact herein is the positive impact we found it is more appropriate to change the measure to "Improve Business Process."

THE PROPOSED MODEL

The above mapping and analysis process yielded the preliminary model of this study as shown in Figure 8.

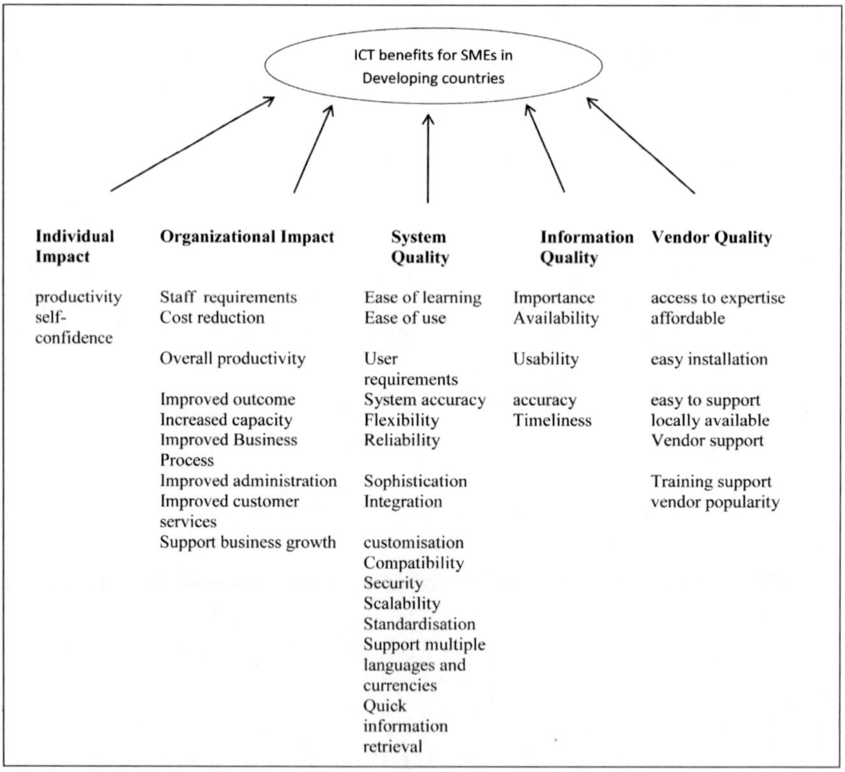

2	9	15	5	8
Total 39				

Figure 8. The proposed model.

The model reflects the characteristics of SMEs and developing countries in many aspects. The model is advantageous given its simplicity which mirrors the ICT skills and limitations of SMEs in developing countries.

CONCLUSION

This chapter aimed to develop a measurement success model for ICT in SMEs in developing countries. It reviewed the literature for the characteristics of SMEs in developing countries. The model derived using a content analysis of 20 published case studies from practice. Analysing the customer quotations, the study identified 340 benefits which were then synthesized into 64 non-overlapping benefits.

The mapping of benefits to the IS-Impact measurement model is the conceptual contribution of this research. Many benefits cannot be covered in existing IS-Impact models, which demonstrates that a new benefits measurement model for ICT in SMEs in developing countries is necessary. Based on characteristics of both SMEs and developing countries, plus the mapping results, the study consolidates a preliminary measurement model for ICT in SMEs in developing countries. The model consists of 5 dimensions with 39 benefit measures. Future research will involve the testing of the proposed model using a quantitative survey.

Despite its contribution, the study has various limitations. Firstly, as the study draws from a single measurement model (IS-Impact), the list of measures is not an exhaustive one. Other models and measures should be included for further consideration. Secondary data sources in vendor-published successful stories that were used in this study can reflect biases by vendors who may overstate the successes and benefits of their products and perhaps avoid their failures.

However, given that the focus of the study is to develop a benefits model that can be later employed to gauge the 'level of success' in a number of organizations, the use of such success stories do not possess a great issue. Furthermore, combining those cases with a primary data source such as survey or case study will create a more realistic view of the findings. Lastly, drawing from a larger number of respondents would naturally convey reliability to the findings and create the possibility for further comparisons.

Nevertheless, this study is instrumental as a necessary first step in equipping SMEs in developing countries will be useful framework to assess benefits of ICT.

REFERENCES

Alam, S. S., & Noor, M. K. M. (2009). ICT adoption in small and medium enterprises: an empirical evidence of service sectors in Malaysia. *International Journal of Business and Management, 4*(2), 112.

Alghamdi, I., Goodwin, R., & Rampersad, G. (2011). A Suggested E-Government Framework for Assessing Organizational E-Readiness in Developing Countries *Informatics Engineering and Information Science* (pp. 479-498).

Alghamdi, I. A., Goodwin, R., & Rampersad, G. (2011). E-Government Readiness Assessment for Government Organizations in Developing Countries. *Computer and Information Science, 4*(3), 1913-1938.

Apulu, I., & Latham, A. (2010). Benefits of information and communication technology in small and medium sized enterprises: a case study of a Nigerian SME. *UK Academy for Information Systems Conference Proceedings 2010*, 7.

Ashrafi, R., & Murtaza, M. (2008). Use and Impact of ICT on SMEs in Oman. *The Electronic Journal of Information Systems Evaluation, 11*(3), 125-138.

Avgerou, C. (2008). Information systems in developing countries: a critical research review. *Journal of Information Technology 23*(3), 133-146.

Ballantine, J., Bonner, M., Levy, M., Martin, A., Munro, I., & Powell, P. L. (1996). The 3-D model of information systems success: the search for the dependent variable continues. *Information Resources Management Journal, 9*(4), 5-14.

Berisha-Namani, M. (2009). *The role of information technology in small and medium sized enterprises in Kosova.* Paper presented at the Fulbright Academy 2009 Conference Small Places Can Change the World.

Blau, P. M. (1970). A Formal Theory of Differentiation in Organizations. *American Sociological Review, 35*(2), 201-218.

Blili, S., & Raymond, L. (1993). Information technology: Threats and opportunities for small and medium-sized enterprises. *International Journal of Information Management, 13*(6), 439-448.

Bohórquez, V., & Esteves, J. (2008). Analyzing SMEs Size as a Moderator of ERP Impact in SMEs Productivity *Communications of the IIMA, 8*(3), 67-II.

Burton-Jones, A., & Gallivan, M. J. (2007). Toward a deeper understanding of system usage in organizations: a multilevel perspective. *MIS Quarterly, 31*(4), 657-679.

Burton-Jones, A., & Straub, D. W. (2006). Reconceptualizing System Usage: An Approach and Empirical Test. *Information Systems Research, 17*(3), 228-246.

Chen, J. C. (1993). The impact of microcomputers on small businesses:England 10 years later. *Journal of Small Business Management, 31*(3), 96-102.

Chinn, M. D., & Fairlie, R. W. (2006). ICT use in the developing world: an analysis of differences in computer and internet penetration: National Bureau of Economic Research.

Consoli, D. (2012). Literature Analysis on Determinant Factors and the Impact of ICT in SMEs. *Procedia - Social and Behavioral Sciences, 62*(0), 93-97.

Cragg, P. B., & King, M. (1993). Small-Firm Computing: Motivators and Inhibitors. *MIS Quarterly, 17*(1), 47-60.

Cragg, P. B., & Zinatelli, N. (1995). The evolution of information systems in small firms. *Information & Management, 29*(1), 1-8.

DeLone, W. H., & McLean, E. R. (1992). Information Systems Success: The Quest for the Dependent Variable. *Information Systems Research, 3*(1), 60-95.

DeLone, W. H., & McLean, E. R. (2003). The DeLone and McLean model of information systems success: a ten-year update. *Journal of Management Information Systems, 19*(4), 9-30.

Delone, W. H., & Mclean, E. R. (2004). Measuring e-Commerce Success: Applying the DeLone & McLean Information Systems Success Model. *International Journal of Electronic Commerce 9*(1), 31-47.

DelVecchio, M. (1994). Retooling the staff along with the system. *Bests Review, 94*(11), 82-83.

Ein-Dor, P., & Segev, E. (1978). Organizational Context and the Success of Management Information Systems. *Management Science, 24*(10), 1064-1077.

Fathian, M., Akhavan, P., & Hoorali, M. (2008). E-readiness assessment of non-profit ICT SMEs in a developing country: The case of Iran. *Technovation, 28*(9), 578-590.

Gable, G. G., Sedera, D., & Chan, T. Z. (2008). Re-conceptualizing information system success: The IS-Impact measurement model. *Journal of the Association for Information Systems, 9*(7), 377-408.

Gooding, R. Z., & Iii, J. A. W. (1985). A Meta-Analytic Review of the Relationship between Size and Performance: The Productivity and Efficiency of Organizations and Their Subunits. *Administrative Science Quarterly, 30*(4), 462-481.

Grama, A., & Fotache, D. (2007). ICT and ERP applications challenges in Romanian SMEs. *Analele Stiintifice ale Universitatii Alexandru Ioan Cuza Iasi, 14*.

Grazzi, M., & Vergara, S. (2012). ICT in developing countries: Are language barriers relevant? Evidence from Paraguay. *Information Economics and Policy, 24*(2), 161-171.

Guy, G. G., & Harold Joseph, H. (1993). *Information Security in the Small Systems Context: A Framework for Understanding*. Paper presented at the Proceedings of the IFIP TC11, Ninth International Conference on Information Security: Computer Security.

Hawari, A. a., & Heeks, R. (2010). Explaining ERP failure in a developing country: a Jordanian case study. *Journal of Enterprise Information Management, 23*(2), 135-160.

Heeks, R. (2002). Information Systems and Developing Countries: Failure, Success, and Local Improvisations. *The Information Society, 18*(2), 101-112.

Holzinger, A. G., & Hotch, R., . (1993). Small firms usage patterns. *Nations Business, 81*(8), 39-42.

Irefin, I., Abdul-Azeez, I., & Tijani, A. (2012). An Investigative Study of the Factors Affecting the Adoption of Information and Communication Technology in Small and Medium Scale Enterprises in Nigeria. *Australian Journal of Business and Management Research Vol, 2*(02), 01-09.

Jennex, M. E., & Olfman, L. (2004). *Assessing knowledge management success/effectiveness models.* Paper presented at the System Sciences, 2004. Proceedings of the 37th Annual Hawaii International Conference

Kale, P., Banwait, S., & Laroiya, S. (2010). Performance evaluation of ERP implementation in Indian SMEs. *Journal of Manufacturing Technology Management, 21*(6), 758-780.

Kartiwi, M., & MacGregor, R. (2007). Electronic Commerce Adoption Barriers in Small to Medium-Sized Enterprises (SMEs) in Developed and Developing Countries: A Cross-Country Comparison. *Journal of Electronic Commerce in Organizations,, 5*(3), 35-51.

Khan, E. H., & Khan, G. M. (1992). Microcomputers and small businesses in Bahrain. *Industrial Management and Data Systems, 92* (6), 24-28.

Kimberly, J. R. (1976). Organizational Size and the Structuralist Perspective: A Review, Critique, and Proposal. *Administrative Science Quarterly, 21*(4), 571-597.

Kulkarni, U. R., Ravindran, S., & Freeze, R. (2007). A Knowledge Management Success Model: Theoretical Development and Empirical

Validation *Journal of Management Information Systems 23*(No. 3), pp. 309 - 347.

Lee, H., Kim, J., & Kim, J. (2007). Determinants of success for application service provider: An empirical test in small businesses. *International Journal of Human-Computer Studies, 65*(9), 796-815.

Levy, M., & Powell, P. (2000). Information systems strategy for small and medium sized enterprises: an organisational perspective. *The Journal of Strategic Information Systems, 9*(1), 63-84.

Lin, C. Y.-Y. (1998). Success Factors of Small- and Medium-Sized Enterprises in Taiwan: An Analysis of Cases. *Journal of Small Business Management, 36*(4), 43-56.

MacGregor, R. C., & Vrazalic, L. (2006). E-Commerce Adoption Barriers in Small Businesses and the Differential Effects of Gender. *Journal of Electronic Commerce in Organizations, 4(2)*, 1-24.

Manochehri, N., Al-Esmail, R., & Ashrafi, R. (2012). Examining the Impact of Information and Communication Technologies (ICT) on Enterprise Practices: A Preliminary Perspective from Qatar. *The Electronic Journal of Information Systems in Developing Countries, 51.*

Mc Cartan-Quinn, D., & Carson, D. (2003). Issues which Impact upon Marketing in the Small Firm. *Small Business Economics, 21*(2), 201-213.

Modimogale, L., & Kroeze, J. H. (2011). The Role of ICT within Small and Medium Enterprises in Gauteng. *Communications 2011.*

Molla, A., & Licker, P. S. (2001). E-commerce systems success: An attempt to extend and respecify the DeLone and McLean model of IS success. *Journal of Electronic Commerce Research, 2, 4 (2001), 1-11.*

Myers, B. L., Kappelman, L. A., & Prybutok, V. R. (1998). A comprehensive model for assessing the quality and productivity of the information systems function: toward a theory for information systems assessment *Information systems success measurement* (pp. 94-121). North Texas University, Information Systems Research: IGI Publishing.

Ndiege, J. R. A., Wayi, N., & Herselman, M. E. (2012). Quality Assessment of Information Systems in SMEs: A Study of Eldoret Town in Kenya. *The Electronic Journal of Information Systems in Developing Countries, 51.*

Osterwalder, A. (2002). ICT in developing countries, A cross-sectoral snapshot. *HEC, University of Lausanne*, 1-13.

Patel, F., Sooknanan, P., Rampersad, G., & Mundkur, A. (2012). *Information Technology, Development, and Social Change* (Vol. 32). New York: Routledge.

Petter, S., DeLone, W., & McLean, E. (2008). Measuring information systems success: models, dimensions, measures, and interrelationships. [Review]. *European Journal of Information Systems, 17*(3), 236-263. doi: 10.1057/ejis.2008.15

Pitt, L. F., Watson, R. T., & Kavan, C. B. (1995). Service Quality: A Measure of Information Systems Effectiveness. *MIS Quarterly, 19*(2), 173-187.

Plewa, C., Troshani, I., Francis, A., & Rampersad, G. (2012). Technology adoption and performance impact in innovation domains. *Industrial Management & Data Systems, 112*(5), 748-765.

Plewa, C., Troshani, I., & Rampersad, G. (2011). Organisational adoption of e-business: the case of an innovation management tool at a university and technology transfer office. *International Journal of Networking and Virtual Organisations, 9*(3), 265-282.

Rampersad, G., Plewa, C., & Troshani, I. (2012). Investigating the use of information technology in managing innovation: A case study from a university technology transfer office. *Journal of Engineering and Technology Management, 29*(1), 3-21.

Sabherwal, R., Jeyaraj, A., & Chowa, C. (2006). Information System Success: Individual and Organizational Determinants. *Management Science, 52*(12), 1849-1864.

Saunders, C. S., & Jones, J. W. (1992). Measuring performance of the information systems function. *Journal of Management Information Systems, 8*(4), 63-82.

Seddon, P., & Kiew, M.-Y. (1996). A Partial Test and Development of Delone and Mclean's Model of IS Success. *Australasian Journal of Information Systems, 4*(1).

Seddon, P. B. (1997). A Respecification and Extension of the DeLone and McLean Model of IS Success. *Information Systems Research, 8*(3), 240-253. doi: 10.1287/isre.8.3.240

Seddon, P. B., Staples, S., Patnayakuni, R., & Bowtell, M. (1999). Dimensions of information systems success. *Commun. AIS, 2*(3es), 5.

Sedera, D., & Gable, G. G. (2004). *A factor and structural equation analysis of the enterprise systems success measurement model.* Paper presented at the Proceedings of the 25th. ICIS, Washington DC, pp.449–464.

Seibert, M. J., III. (2004). *The identification of strategic management counseling competencies essential for the Small Business and Technology Development Center: A modified Delphi study.* Ed.D., North Carolina State University, United States - North Carolina.

Shang, S., & Seddon, P. B. (2002). Assessing And Managing The Benefits Of Enterprise Systems: The Business Manager's Perspective. *Information Systems Journal, 12(4), 271-299.*

Skoko, H. (2012). Influencing Factors Model of Information and Communication Technology (ICT) in Saudi Arabian Small and Medium Enterprises (SMEs). *The GCC Economies*, 229-234.

Skoko, H., & Ceric, A. (2010). *Study on Information and Communication Technology (ICT) Models of Adoption and Use in the Kingdom of Saudi Arabian SMEs.* Paper presented at the The 6th International Conference on Social and Organizational Informatics and Cybernetics: SOIC 2010 in the context of The 4th International Multi-Conference on Society, Cybernetics and Informatics: IMSCI 2010, Orlando, FL USA.

Snider, B., da Silveira, G. J. C., & Balakrishnan, J. (2009). ERP implementation at SMEs: analysis of five Canadian cases. *International Journal of Operations & Production Management, 29*(1/2), 4-29.

Vrgovic, P., Glassman, B., Walton, A., & Vidicki, P. (2012). Open innovation for SMEs in developing countries-an intermediated communication network model for collaboration beyond obstacles (revision). *INNOVATION: Management, Policy & Practice*(0), 587-614.

Wei, K. S., Loong, A. C., Leong, Y.-M., & Ooi, K.-B. (2009). *Measuring ERP system success: a respecification of the Delone and McLean's IS success model.* Paper presented at the Symposium on progress in information & communication technology.

Wu, J.-H., & Wang, Y.-M. (2006). Measuring KMS success: A respecification of the DeLone and McLean's model. *Information & Management, 43*(6), 728-739.

Yap, C., Soh, C., & Raman, K. (1992). Information systems success factors in small business. *Omega, 20*(5-6), 597-609.

Zhu, K., & Kraemer, K. L. (2005). Post-Adoption Variations in Usage and Value of E-Business by Organizations: Cross-Country Evidence from the Retail Industry. *Information Systems Research, 16*(1), 61-84.

In: Technology Innovation Leadership … ISBN: 978-1-63117-388-2
Editors: G. Rampersad and F. Patel

Chapter 9

ADVOCATING A HOLISTIC FRAMEWORK FOR INNOVATION LEADERSHIP AS A SOCIAL RESPONSIBILITY IMPERATIVE

Fay Patel*

Bond University, Gold Coast, Queensland Australia

ABSTRACT

Over recent decades, the literature is exhausted with innovation diffusion models and comparative studies with a concentration on the pace of diffusion and the volume of population reach. Capacity building and technology transfer feature as marginal components to develop and upskill the recipients of innovation diffusion. A proactive holistic framework to lead innovation diffusion remains absent. In this chapter, the author advocates a holistic framework for innovation leadership as a social responsibility imperative.

INTRODUCTION

Aubert (2005, p.6) contends that 'the promotion of innovation, in particular technological innovation, in developing countries is becoming a fashionable subject'. He claims that there is an emerging recognition [in

* E-mail: dr.fay.patel@gmail.com.

western circles] (author insert) that traditional economic development policies and practices had 'limits for promoting sustainable growth in the developing world'. Of particular note, is Aubert's contention that 'policies focusing on modernization, in the sense of building infrastruture and institutions with a more interventionist government' did not meet the envisaged expectations [of the exporters of innovation to developing countries] (author insert). As mentioned in an earlier chapter, Aubert is emphatic that *innovation* refers to both technological and 'non-technological forms of innovation' that may include the introduction of new ideas, products and services to drive economic agendas. There is an admission that the highly sophisticated innovation policy development and application (p.6) 'in developed/OECD countries, ... is not directly applicable to developing countries because of the nature of the challenges that latter are facing'.

Nearly 50 years after the great surge of innovation exportation (under the guise of modernization) from the west (see Aubert's reference to the Charpie Report in the USA in 1964) to the developing world Aubert (p.6) recognises that 'developing countries face genuine obstacles to innovation'. It is encouraging to note his advocacy for a critical reflection of ' innovation approaches [that can be] adapted to the needs and possibilities of developing countries. Aubert uses the labels 'developing world' and 'developing countries' interchangeably to refer to those countries in which obstacles to innovation 'derive from inappropriate business and governance climates and insufficient education'. In this chapter the author also uses the term 'developing communities' (Patel et al, 2012) to refer to those communities that are less advantaged through poverty and economic hardship along with politically challenging contexts. As noted in Aubert (2005) and Gaventa (2004), the realization that innovation policy schemes developed in the west cannot be modeled within a one size fits all approach. This is a remarkable shift in the innovation diffusion and modernization paradigm nearly half a century later. It is a positive, proactive stance and differs widely from that of innovation diffusion thinking and implementation within the traditional export model espoused by Schramm (1936), Lerner (1958), and Rogers' (2005).

According to Aubert (2005, p.7), 'one single model of development ... would perpetuate inequalities and the status quo' and so he advocates an alternate approach for the 'gradual transformation of developing countries'. Aubert's (p.7) alternate approach to innovation diffusion and economic development focuses on the 'promotion of innovation' through 'building upon resources and capabilities available in countries at their level of development and in taking due account of their specificities including their conditions of governance'. Gaventa (2004) expresses similar sentiment. In sum, both Aubert (2005) and Gaventa (2004) are affirming a long held belief among developing communities and some international development professionals: there is a moral imperative to provide a holistic framework for innovation that aligns with developing community needs and perspectives. Innovation should be framed within an enabling paradigm and should be responsive to empowered stakeholders so that they can action the necessary changes for sustainable living. An enabling paradigm is one that:

- Communicates (in an open and transparent manner) about the innovation within a rights and responsibility framework
- Encourages capability building and provides opportunity and mechanism for building capability
- Clarifies the roles and responsibilities of stakeholders in the innovation
- Responds to empowered stakeholder requests to action change
- Encourages modifications to legislation to include and diversify governance policies and practices so that there is equity in the distribution of services, funds and opportunities for growth and development

Both Aubert (2005) and Gaventa (2004) explore alternate approaches to innovation and make critical recommendations that place developing communities and innovation leaders at the centre of the innovation instead of at the margins (where they have been for over half a century).

Aubert's contention that promoting innovation from the perspective of developing communities is an imperative and Gaventa's (2004, p.1) observation that 'localism' (local community involvement) drives their political economic agendas are worthy of note in this discussion. Furthermore, Gaventa's assertion that the UK can learn important lessons from their international counterparts including the developing world is significant reshaping the innovation diffusion paradigm into a future oriented, collaborative partnership among glocal (merger of local and global) communities. The innovation paradigm of the last fifty years clearly did not acknowledge local communities or developing communities as owners and shapers of innovation initiatives. In traditional models, innovation was externally driven and controlled with little or no recognition or acknowledgement of the indigenous knowledge and socio-cultural contributions of local communities as partners in international development. In the traditional paradigm, local communities were selectively recruited to support the externally driven innovation through 'change agent' and 'community leader' strategies. Returning to the discussion on Aubert and Gaventa's shared perspective that developing communities should be the first point of consultation within the 21st century innovation framework, the next section briefly reviews common threads in their discussion. Further, in identifying these commonalities, the discussion also highlights recommended strategies in upholding a socially responsible imperative.

Ironically, Aubert (USA based) and Gaventa (UK based) are suggesting that the west can learn lessons (in gradual transformational development, participatory democracy, and enabling communities for sustainable futures) from those countries that have been labelled 'developing' as a result of their deficits in business, education and governance.

COMMONALITIES IN APPROACH TO COMMUNITY EMPOWERMENT

Commonalities in the Aubert and Gaventa studies are reviewed in this section. Aubert's (2005) study on promoting an innovation conceptual framework in developing countries and Gaventa's study (2004) on citizen involvement in neighbourhood renewal and local governance collectively promote an empowering paradigm to support developing communities in innovation leadership. Examples of a common approach and shared sentiments to enhance sustainable developing community are listed in the table below.

Table 1. Commonalities in approach to community empowerment

Gaventa 2004	Aubert 2005
A one-size-fits-all approach will not work (p. 2)	One size does not fit all (p.7)
Strengthen the legal or statutory provisions which enable participation (p.2)	Judicious adjustments in the broad regulatory framework (p.8)
Build upon the diversity of local understandings of leadership (p.2)	Recognition that indigenous knowledge is an asset (p.13) to be integrated and mobilizing local communities (p.20)
'work both sides of the equation', simultaneously focusing both on community empowerment and supporting the capacity of local officials and civil servants to understand, and respond to that empowerment; (p.2)	Efficient support for innovators (p.17) delivered in an integrated and coherent way with a maximum degree of flexibility (p.20)

Aubert and Gaventa advocate for holistic frameworks that create opportunities for empowerment and that respond favourably to the creative application of the innovation among empowered citizens. Rampersad (2012, p.24) contends that 'a holistic approach is required to empower communities by providing a model for diffusing innovation that is relevant to developing communities'.

Proactive Holistic Framework to Lead Innovation: A Social Responsibility Imperative

Proactive holistic approaches to innovation diffusion support modifications in governance regulations to include equity, inclusivity and diversity in leading innovation. *A proactive approach enables innovation leaders to lead innovation* through ongoing commitment to equity, expanding opportunities for innovative solutions, and responding favourably and appropriately to the creative integration and blend of indigenous knowledge and innovations. Proactive approaches enable stakeholders to capture multiple perspectives in innovation exploration and look to the future with optimism. *Leading innovation* is different from managing innovation in the following ways: *to lead innovation one requires vision and an open mindset to explore the possibilities*; *to lead innovation it is important to facilitate creativity, promote opportunities, and to be responsive to creative, context-driven solutions.* Managing innovation refers to the bureaucratic 'watchdog' and compliance model of assuring that prescriptive process requirements are met. Proactive holistic frameworks are aligned with social responsibility commitment and are visible through socially responsible actions. The following are some examples of socially responsible actions that should be built into proactive holistic frameworks for innovation diffusion among developing communities. Socially responsible actions include: proactively establishing a holistic framework for innovation; creating spaces in governance for enabling citizens; redistributing and allocating funds to support skills development; establishing a vigilance team to supervise appropriate use of funds; the making available personal and professional development initiatives to utilize the innovation effectively; and responding to creative applications of the innovation proposed by empowered citizens. Innovation diffusion should be embedded within a proactive, holistic framework that is enabling and responsive so that innovation leaders can support communities in building sustainable futures.

As noted in the preceding discussion, a comprehensive holistic framework is advocated as a social responsibility imperative. Such an holistic framework is concerned with issues of human capability and ongoing dialogue as an essential component of innovation diffusion. Within a socially responsible and accountable framework, innovation diffusion is an interactive, negotiated space from inception to implementation through ongoing critical review of the benefits (and/or harm) of the innovation diffusion. An interactive model of innovation diffusion (Patel, 2009, p.107) subscribes to a 'consultative, open communication context from the pre-implementation to the post-implementation phase' among the organisational management and employees. Further, (p.108) 'it is a multilevel diffusion model that aims at monitoring and assessing the diffusion across three phases.' In the first phase (pre-implementation and consultation phase), consultations are held among employees through innovation diffusion forums to seek consensus and once consensus is reached a human capability development program is established to provide innovation training. In the second phase (implementation and utilization and satisfaction phase), organisational communication policy is developed and implemented, training and development is continued. In the next phase (post-implementation and assessment phase), innovation use and satisfaction, training program effectiveness and degree of diffusion within the organisation is assessed, non-users and infrequent users are given incentives to use the innovation, and the diffusion policy and practice is modified to enhance effective use of the innovation. Within an interactive innovation model, both product and process are viewed as complimentary components of innovation.

Emphasis in the holistic framework is on leading innovation for sustainability from within which means that enabling strategies are developed in policy and practice to ensure that local communities can access, use and creatively apply and blend the innovation within a socio-cultural and indigenous knowledge framework that is sustainable. Strategies to ensure sustainability include the following:

- Aligning innovation design and implementation with socio-cultural and indigenous value and belief systems and establishing 'appropriate incentives' (Aubert, 2005, p.28)
- Making an effort to ensure that innovations are user friendly

- Innovation users are given the opportunity to develop adequate capacity to use innovations effectively
- Technical glitches are fixed without delay so that barriers to smooth operations are removed
- All components of the innovation (infrastructure, hardware, software and human capability) are monitored for quality and effectiveness
- Evaluation of the innovation experience in related to enhancement of quality of life (Aubert, 2005, p.30)
- Macro and micro level considerations of innovations should be reviewed frequently with the intention to improve operations (Aubert, 2005, p.30 and p.34)

Nurturing Human Capital and Building Human Capability

Nurturing human capital and building adequate capacity requires a sensitive, human response to human development within the holistic framework of innovation diffusion. As noted in Patel et al (2012, p.20), 'past innovations and development initiatives did not place people first, and so the human component was absent'. The 21st century innovation paradigm is human development oriented with a focus on building human capability for sustainable futures. This paradigm is concerned with the psychological and physical well being of developing communities within an enabling paradigm and one that requires adequate capability building and skill development among developing communities. Mayer (2001, p.34) contends ' that an important determinant of the benefits which developing countries can reap from globalization is whether they can ignite a simultaneous increase of technology diffusion and the skill level of the domestic labour force'. In line with Auberg and Gaventa's sentiments, Mayer asserts that a degree of flexibility in government policy allowing modifications 'to sustain incentives for both human capital formation and a reduction in the cost of technology adoption' are desirable. Again such flexibility is supported in a proactive and responsive innovation approach. More importantly, Mayer claims that 'investment in human capital alone would lead to diminishing returns of skill accumulation, while increased technology transfer alone is unlikely to be enduring and might have negative developmental effects from rising income inequality'. As noted in the discussion from Aubert, Gaventa and Ockwell's perspectives, innovation begins with developing communities as the primary drivers of change, respectfully acknowledging their socio-cultural perspectives and indigenous knowledge. The 'needs and wants' (what developing communities need and what they want in order to survive in the current and future socio-economic, political and environmental global sphere) of developing communities is the starting point for exploring suitable innovation solutions or e-solutions that enable communities through the use of electronic media and information communication technologies. Again, it is encouraging to note the emergence of alternate models of international development and innovation diffusion. Traditional linear models of innovation in the past adhered to dogmatic policy frameworks that supported export (Patel, 2012, pp.17 -19) of innovation diffusion from western contexts to developing communities. Past models were not aligned with social responsibility ethics in innovation.

Respecting Socio-Cultural Practice and Indigenous Knowledge as Cultural Wealth

Western hegemonic models of innovation diffusion and international development in the traditional paradigm were framed in ethnocentric contexts that regarded developing communities as the passive recipients of economic development initiatives from the west. Further, the indigenous knowledge, cultural values and historical legacies of developing communities were 'devalued and dismissed as significant contributions' (Patel, 2012, p.19). It is encouraging to note the Aubert (2005, p.28) refers to 'indigenous knowledge as a specific wealth'. Aubert (2005), Gaventa (2004) and Ockwell (2010) subscribe to a future oriented innovation model that respects the experiences, knowledge and cultural values of local communities. Ockwell et al (2010, p.7) assert that 'Indigenous eco-innovation capabilities are essential to facilitating both the diffusion of existing eco-innovations within developing countries and sustainable economic development based on the adoption, adaption and development of environmentally sound technologies that fit with the bespoke conditions faced by developing countries'. This 21st century innovation perspective recognises indigenous knowledge and capability among developing communities as essential for sustainable development and has respects 'the importance of engaging 'other' worldviews that are embedded within unique but valuable cultural knowledge'(Patel et al, 2012, p. 141)

Aubert (p.28) is cognisant of the imperative to respect 'socio-cultural differences between the developed world and the developing one' and he claims that 'what is at stake is the respect of the socio-cultural foundations of countries in such a way that they effectively blend traditional ways of functioning with modern approaches facilitating their integration into modernity and the global economy'. Within an enabling innovation paradigm, developing communities will not only shape their collective destinies but will also impact their inevitable shared futures with the developed western world.

SUSTAINABLE INNOVATION POLICY AND PRACTICE

Aubert (2005), Gaventa (2004) and Ockwell (2010) offer a wide range of sustainable innovation policies and practices to lead innovations and enable developing communities to set their own development agendas. Gradual transformational development that is suited to the pace of developing community needs will ensure a harmonious and effective shift to sustainable futures.

As noted in the discussion, recent trends are pointing toward the development of sustainable policy and practice that is embedded within a social responsibility framework. Sustainable innovation policy and practice requires a holistic framework within which all stakeholders are held responsibility and accountable for their actions and non-actions. Social responsibility frames innovation within a rights and responsibilities context and

- Demands that enabling mechanisms through legislated policy and practice are developed to empower communities to action change (Gaventa, 2004, p.25)
- Creates accountability measures
- Ensures that stakeholders are not exploited and not short-changed on the promises made to develop their full potential and capability to embrace the innovation
- Monitors funds against misuse
- Establishes mechanisms to monitor diffusion of innovations initiatives within reasonable time frames
- Identifies roadblocks to effective innovation diffusion
- Supports innovation leaders
- Monitoring human capability development
- Connects the different innovation components

- Ensures transparent communication and upholds rights to information
- Promotes an interactive approach to diffusion of innovations (Patel, 2009, pp 107 -109)
- Establishes an ongoing evaluation mechanism to enhance and modify innovation (Aubert, 2005, p.33)

Gaventa (2004, p.25) summarizes the key lessons from international contexts that are an imperative for sustainable innovation frameworks, as noted below.

- 'The importance of enabling legal and statutory provisions: *Having legislation in place legitimates participation as a right, not just as an 'invitation' for consultation'.'(p.25)*
- 'Importance of local and regional contexts: *Consideration of multiple strategies, flexibility in approaches, diversity in measuring success, and establishing pre-conditions for participation, where they are absent. (p.25)*
- 'Working both sides of the equation – strengthening community voice'(p.26): *Providing spaces for participant voice and mechanisms in place to be responsive to those voices is important. Community skill development is recommended in addition to individual skill development.*
- 'Working both sides of the equation – strengthening government response'(p.27): *'...government staff and councillors have been trained to act for the community. Changing to act with the community requires new attitudes and behaviours'.*
- 'Clarifying the identities and accountabilities underlying representative processes'(p.28): *Understanding who represents what and how they are chosen as representatives of the different communities is important*
- 'Improving incentives for engagement and quality representation'(pp.28-29): *When results and actioned change is visible this provides incentive for quality participation*
- 'Garnering support from non-government allies' (p.29): *Social movements give greater impetus to participatory government approaches*
- 'Naming and addressing power relationships'(p.30): *'Participatory processes must engage with and change power relationships'*

- 'Taking time and going slow' (p.30): *Perhaps, the most important lesson of all is that 'new attitudes, new forms of trust and collaboration, new skills and*
  capacities, new models of leadership and power sharing - all take time to develop and to grow'.

It is not the pace at which the innovation is diffused but the quality of that innovation that must take precedent in participatory innovation approaches.

CONCLUSION

A holistic framework for innovation leadership as a social responsible imperative is contingent upon establishing various conditions for enabling innovation leaders and developing communities so that they can take collective responsibility for designing sustainable futures.

Evident in the preceding discussion is Mayer (2001), Gaventa (2004), Aubert (2005) and Ockwell's (2010) commitment to 'a socially responsible, just, and sustainable development perspective' (Patel, 2012, pp.16 -23). This suggests that such a proactive, future oriented innovation paradigm has begun to embrace social responsibility as an imperative. Nussbaum (2011, citing Young in the Foreword - xiii) claims that the 'demanding standard to assess responsibility' requires that '... a responsible person tries to deliberate about actions before acting, makes choices that seem to be the best for all affected, and worries about how the consequences of his or her actions may adversely affect others' Furthermore, Nussbaum (xvii) contends that people 'who have greater influence have greater responsibility'.

Gaventa (2004) has clearly identified roles, responsibilities, and accountabilities within a socially responsible and just innovation framework by outlining those conditions (and pre-conditions) that should be in place in order to enable communities to act in their own best interest. For example, Gaventa's assertion that creating spaces for 'voice' is not demonstrative of adequate social responsibility and justice unless it is accompanied by established policy and practice that is responsive to the empowered voice. This illustrates the depth and breadth within which a socially responsible, just and sustainable innovation paradigm can be effective. It is only effective if pre-conditions are set to enable communities and post-conditions are in place to positively embrace and support the innovations that emerge within empowered communities who are inspired to explore context-relevant innovations.

As with social responsibility so too with justice: 'one cannot advocate justice if one has not taken responsibility for providing an environment in which fellow human beings are treated in a just manner' (Patel, 2012, p.139). For example, promising great strides in socio-economic and political advancement through innovation should be firmly embedded within an environment in which such advancement is visibly supported by incentives and rewards across the complex dimensions of national, industry, organisational and individual dimensions, as noted by Rampersad (2012, p.25). It is within that paradigm that responsibility is executed and justice is 'served if one uses the means one has to make a difference to the quality of life of those who are oppressed' (Patel, 2012, p.139), regardless of the form of that oppression. As noted by Danielle Allen (ix), '... Young takes justice to consist of the social and institutional conditions necessary for creating nondomination and nonoppression, where the latter means the achievement of human flourishing, for all members of society'. The emerging 21st century innovation paradigm together with a visible commitment to social responsibility (and justice) as an imperative in current innovation literature holds promise for building sustainable communities who will continue to shape their collective destinies.

REFERENCES

Aubert, J-E (2005) *Promoting Innovation in Developing Countries: A conceptual framework World Bank Institute Contributions by Kuznetzov,* Y., Whyte, R., & Quirke, A.

Gaventa, J. (2004). *Representation, community leadership, and participation: Citizen involvement in neighbourhood renewal and local governance Institute of Development Studies Prepared for the Neighbourhood Renewal Unit,* Office of the Deputy Prime Minister, UK.

Lerner, D. (1958). *The passing of traditional society: Modernizing the Middle East New York,* USA: Free Press.

Mayer, Jorg. (2001). *Technology diffusion, human capital and economic growth in developing countries No. 154 United Nations Conference on Trade and Development* (UNCAD) Discussion Papers.

Ockwell, D., Watson, J., Mallet, A., Haum.R., MacKerron, G. & Verbeken, A-M. (2010), *"Enhancing Developing Country Access to Eco-Innovation: The Case of Technology Transfer and Climate Change in a Post-2012 Policy.*

Framework", *OECD Environment Working Papers*, No. 12, OECD Publishing.

http://dx.doi.org/10.1787/5kmfplm8xxf5-en.

Patel, F., Sooknanan, P., Rampersad, G., and Mundkur, A. (2012) *Information Technology, Development, and Social Change*, New York: Routledge. ISBN 13: 9780415502689.

Patel, F. (2012). Reframing the diffusion of innovations and international development within a socially responsible, just and sustainable development perspective In Patel, F., Sooknanan, P., Rampersad, G., & Mundkur, A. (2012). *Information Technology, Development, and Social Change*. New York: Routledge.

Patel, Feiziya .(2009). *Role of Attitudes and Perceptions in Diffusion of Innovations Implementation of the Internet and Email in South Africa Saarbrucken,* Germany: VDM Verlag Dr Muller.

Rogers, E. M. (2005) *Diffusion of innovations*, 5th Edition, New York: Free Press.

Schramm, W. (1964). *Mass media and national development Stanford*, CA: Stanford University Press.

Young, Iris M. (2011). *Responsibility for Justice* N.Y., USA: Oxford University Press Foreword by Marha Nussbaum.

Young, Iris M. (2011b). *Justice and the Politics of Difference USA:* Princeton University Press 2nd Edition Foreword by Danielle Allen.

In: Technology Innovation Leadership … ISBN: 978-1-63117-388-2
Editors: G. Rampersad and F. Patel

Chapter 10

DIGITAL INNOVATION STRATEGIES FOR DEVELOPING COMMUNITIES IN THE NEW CENTURY

Giselle Rampersad[1,*] and Fay Patel[2]
[1]Flinders University Australia
[2]Bond University, Gold Coast, Queensland Australia

ABSTRACT

This chapter reviews the contributions to the book and examines digital innovation strategies for developing communities in the new century. It highlights the encouraging engagement of innovation leaders and developing communities to influence innovation diffusion from within developing community contexts.

In determining digital innovation strategies for communities, it addresses important areas for e-government in not only improving e-government readiness but also in taking steps to develop e-government maturity; e-commerce in addressing unique barriers in developing countries; the key role of social media and ensuring benefits of information communication technology (ICT) for small firms in developing countries, and tools to monitor environmental sustainability and to ensure that there is effective utilization of ICT infrastructure through high-speed broadband.

[*] Corresponding author: giselle.rampersad@flinders.edu.au.

INTRODUCTION

Aubert (2004, p.7) contends 'that that the promotion of innovation should be considered in a gradual manner in building upon resources and capabilities available in countries at their level of development and in taking due account of their specificities including their conditions of governance'. This is a refreshing alternate approach to innovation that is embedded within a developing community perspective. Viewing innovation from within a developing community context takes cognisance of developing community needs and opportunities (p.11) for innovation. Aubert (p.11) identifies three forms of innovation:

- 'local improvements based on the adoption of technologies which are more or less available worldwide or locally ("technology adoption" from a global perspective)'
- 'building up of competitive activities with some adaptation made to existing technologies ("technology adaptation")'
- 'design and production of technologies of a worldwide significance ("technology creation" from a global perspective)'.

Against Aubert's (2004) forms of innovation, the innovation leaders in this book have diversified their innovation approaches according to their community needs and opportunities.

DIGITAL INNOVATION STRATEGIES WITHIN ENABLING INNOVATION FRAMEWORKS

Governments have become increasingly concerned with ensuring that they build knowledge based economies that have the capacity to innovate. In developing communities, the technology systems and infrastructure are usually inadequate. Consequently, leadership is essential to not only recognise these inadequacies but to develop innovative solutions despite these inadequacies. These leaders are key catalysts of change to ensure that communities experience enhanced quality of life and community development. The contributors of the book examined several steps that are essential in developing innovative digital strategies in developing countries. These are highlighted in Figure 1 and discussed in greater detail throughout this chapter.

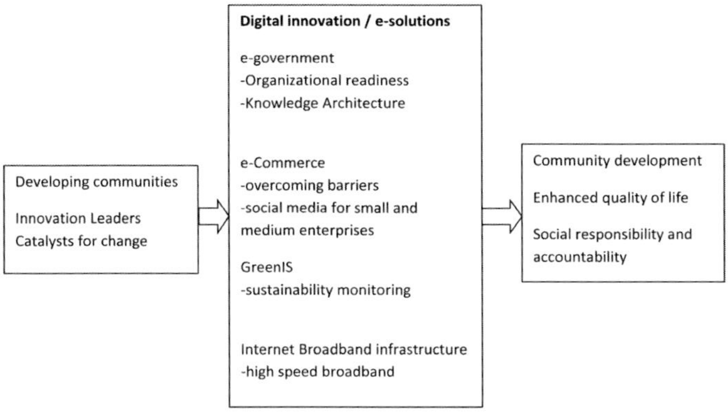

Figure 1. Developing digital innovation strategies within enabling innovation frameworks.

Enabling Innovation Leaders and Developing Communities as Catalysts for Change

In Chapter 1, Patel contends that the 21st century paradigm is an enabling paradigm calling for the recognition of innovation leaders (and developing communities) as catalysts for change from within developing contexts. Enabling strategies are highlighted as proactive, responsive mechanisms to support innovation leaders and their communities in strengthening their resolve to build sustainable futures. Newly emerging innovation leadership from within developing community contexts is taking charge of innovation as a commitment to serve their communities and to enhance their quality of life. Furthermore, there is increasing acknowledgement of the role of innovation leaders within their communities as a necessary 'innovation from within' innovation model that has long been advocated among developing communities.

Evaluate the Readiness of Government Organizations for e-Government

In Chapter 2, Alghamdi discusses the need to employ tools to evaluate the readiness of government organizations for e-government, with a focus on the key factors of the level of e-services offered, the adequacy of ICT infrastructure and the capacity of the workforce in ICT skills. These factors were derived from the Center for International Development (Harvard University), the Asian Pacific Economic Cooperation (APEC) models, and suggested factors by the research team of Yesser which is the national e-government program in Saudi Arabia. His results show that more efforts are fundamentally required to advance e-government in the Saudi Arabian public sector particularly regarding e-services and human skills. In developing appropriate solutions, he recommends i) improving co-ordination and compatibility with National e-government programs, ii) improving co-ordination and collaboration among organizations, iii) increasing the number of e-services provided, v) enhancing ICT infrastructure and connectivity speed internally and with other public organizations, and iv) paying more attention to the attraction, development and retention of talented staff by appropriate remuneration and specialized e-government training.

Improve the Maturity of e-Government

In Chapter 3, Hanafiah and Goodwin discussed strategies for organizations to become more mature in their e-government capacity through improved enterprise architecture. Public organizations become more mature in e-government as they move from mere information dissemination to two-way communication, integration, facilitation of transactions to full participation. The chapter examined approaches to developing maturity through organizational alignment so that the organization's members share a common understanding which in turn improves reliability, responsiveness, assurance and empathy among each other; information availability to find out the usefulness, and accessibility of output from e-Government; and resource portfolio optimization and resource complementary to ensure that there will be efficient infrastructure investment. These factors contribute to improved service quality in the delivery of government services.

Devise Innovative Green Information Solutions for Developing Clean Technology Industries

In Chapter 4, Fradley and Rampersad examined how new legislation coupled with the introduction of a green information system to enfource compliance has lead to the development of a vibrant industry and significant exports around water treatment systems. Employing a community integrated approach to the introduction of sustainability legislation, the chapter investigated the effective use of technology in monitoring water pollution from commercial and residential vessels and in ensuring compliance to sustainability regulation. Key community stakeholders include fishing communities, boat builders and repairers, marine construction workers and equipment suppliers, waste contractors, local government agencies as well as governmental environment and water departments. The chapter provides useful insights for government agencies and buinsesses involved in the development of clean technology innovative industries.

Support Small and Medium Enterprises in Their Use of Social Media

In Chapter 5, Barashi called for more support to be given to small and medium enterprises (SMEs) in their use of social media. Although SMEs are normally time poor, they are the engines of growth and innovation and if used effectively, social media can be a powerful tool for them. Businesses cannot be sustained with traditional marketing tools but SMEs should incorporate social media to engage more with consumers. Social media is not just another marketing channel from face-to-face, post, telephone, to the Internet but it has an interactive component that adds value to the business offering. A significant challenge for SMEs is the lack of knowledge in the measurement and the effective use of SNS.

Another challenge is the lack of expectation of what might be written (comments, complaints, negative posts). Therefore, programs to develop social media skills and to manage engagement with consumers using this medium are instrumental in supporting SMEs in their effective use of social media. Education in assisting SMEs with developing strategies is important to identify key objective as well as how, why and when will social media be used for the benefit of the business.

Implement Solutions with Sensitivity towards Social, Cultural and Economic Factors Impacting Upon e-Commerce Adoption

Technology transfer of Western e-commerce models may not necessary work in developing counties until there is adequate consideration of the unique challenges facing such communities. Developing countries may face social, cultural and economic barriers to e-commerce adoption. In Chapter 6, Alqahtani, Goodwin and de Vries uncovered key issues in Saudi Arabia. Delivery issues are prevalent as street addresses for homes are uncommon. Some goods are not conducive to postage in such extreme heat conditions such as chocolate and electronics. Language barriers preventing understanding of foreign e-commerce websites and rejection of Western advertising such as the use of women in selling products is also prevalent.

More broadly, a lack of infrastructure, internet adoption, legislation and regulations in Saudi Arabia has disabled an environment conducive to e-commerce. The high costs of settling up online businesses there have also been a key hindrance. Therefore, consideration of cultural, social and economic issues is important in e-commerce efforts in such countries.

Implement Monitoring Tools to Ensure That the Community Reaps the Benefits of the Internet

Despite rhetoric about the economic and community benefits of high-speed broadband (the Internet) given billions of dollars invested in many countries, in Chapter 7, Rampersad argues that such benefits are seldom monitored. Monitoring that the Internet benefits are in fact achieved is important in holding governments to account, ensuring the public investment is well allocated and for assessing the progress of government-led initiatives. Although the Internet promises many benefits to the business sector, important factors should be monitored such as the impact of the Internet on productivity, competitiveness and innovation. Teleworking has been a contentious issue. Despite its anticipated benefits on productivity, ironically, ICT firm, Google has recalled their officers back from home to workplace accusing some of skylarking at home. Key issues to be addressed include whether practices and policies have been adjusted to take advantage of teleworking opportunities?

E-education has also been an important selling point of the Internet. However, monitoring must occur on the quality and adequacy of teaching materials for the mode of delivery. Is training synchronous as this will harness the Internet to a greater extent than non-synchronous, non-interactive education? Do students have access to the Internet both at home and educational institution? Is the Internet mode of delivery complementary or substitute for other important modes of delivery for the transfer of tacit knowledge? Does Internet-based learning offered sufficiently equip students with skills in innovation needed in the 21st century?

In addition to e-education, e-health is another key area for the Internet. Important questions arise: Are there opportunities for tele-radiology which involves the downloading of large files? Is narrowband sufficient for tele-diagnosis or is the Internet really necessary? Have barriers been removed that influence the uptake of tele-medicine? Would telemedicine be rolled out in urban communities or mainly rural communities where the impact may be more heavily felt?

The impact of the Internet on community development should also be monitored. Have opportunities to include the socially excluded, aged and disabled been adequately explored? Have online applications been developed to foster a socially vibrant community, by listing community events on websites and mobile applications? Have appropriate policies been put in place to ensure that groups of citizens based on demographics of age, education and income are given equal opportunities to participate in political and community decisions?

To reap the benefits of large investments of broadband infrastructure, it is important that government monitor the actual impact of such infrastructure on economic and community development.

Equip Small Businesses with Tools to Evaluate the Effectiveness of ICT

In Chapter 8, Alshardan, Goodwin and Rampersad, developed a tool to equip small and medium enterprises (SMEs) in developing countries with the ability to evaluable whether ICT services are effective for their organizations given the unique challenges that they face. Although much research has been conducted on the benefits of ICT on both large and small organizations mainly in developed countries, that chapter made an important contribution in devising an appropriate tool for SMEs in developing countries in particular. These SMEs have substantially different characteristics, needs and resources compared to organizations of all sizes in developed countries.

The framework identified five key areas of individual impact, organizational impact, system quality, information quality and vendor quality. This research contributed a vital IS-Impact measurement model for use in developing countries.

ADVOCATING A HOLISTIC FRAMEWORK FOR INNOVATION LEADERSHIP AS A SOCIAL RESPONSIBILITY IMPERATIVE

In Chapter 9, Patel advocates for a holistic framework for innovation leadership and explores innovation approaches that subscribe to a social responsibility (and justice) ethic. Within that perspective, Patel unravels new approaches to innovation that recognise the developing community contexts as significant contributors to transformational change to various socio-economic and political structures. The social responsibility imperative is more significantly embedded within innovation leadership with proponents calling for diversity in innovation approaches and flexibility to embrace innovations through modifications in legislated policy and practice. This is a remarkable transformation with the innovation paradigm itself and far removed from its dogmatic, linear beginnings in the 1950s and 1960s.

Gurstein (2003) maintains that the effective use'approach to information communication technology innovations within local communities requires a diversification and flexibility in the approach to the use of the innovation. He contends that application of innovation that is connected to local needs and providing local communities an opportunity to be creative 'would also give an application the capacity to evolve and respond to local circumstances — constraints and opportunities — as they present themselves.

CONCLUDING REMARKS

The contributors to this book have all demonstrated their commitment to building sustainable communities through an ongoing search for digital innovation strategies to uphold a high standard and quality of life. Socially responsible and just policies and practices are critical to building sustainable futures.

Innovation leaders and developing communities have begun to take their rightful place in the mainstream of the 21st century innovation paradigm. As noted in the literature and in the diverse range of discussion in this volume, innovation leaders and their developing communities are leading change in a way that had been unimagined nearly fifty years ago. Developing communities will determine from within what innovations are necessary, when those innovations are required, and how they will blend indigenous knowledge, cultural wealth and innovative perspectives to shape their futures.

REFERENCES

Aubert, J-E (2004) *Promoting Innovation in Developing Countries: A conceptual framework World Bank Institute Contributions by Kuznetzov,* Y., Whyte, R., & Quirke, A. Retrieved from http://info.worldbank.org/etools/docs/library/137729/0-3097AubertPaper%5B1%5D.pdf.

Gurstein, M. (2003). *Effective use: A community informatics strategy beyond the digital divide by Michael Gurstein First Monday*, Volume 8, Number 12 - 1 December 2003. Retrieved from http://firstmonday.org/ojs/index.php/fm/article/viewArticle/1107/1027".

EDITORS' CONTACT INFORMATION

Dr. Giselle Rampersad,
Senior Lecturer in Innovation and Entrepreneurship
Flinders University
School of Computer Science,
Engineering and Mathematics
GPO Box 2100,
Adelaide 5001 South Australia
Tel: +61 882015746
E-mail: giselle.rampersad@flinders.edu.au

Dr. Fay Patel,
Office of Learning and Teaching
Bond University
Queensland 4229 Australia
Tel: +61 481343877
Email: fpatel@bond.edu.au

INDEX

E

G

H

I

K

L

M

O

P

Q

R

S

U

V

W

Y